Gödel's Way

A personal account by some of the participants in the work going beyond Gödel by finding uncomputability and incompleteness in many areas of continuous and discrete mathematics and theoretical physics.

Gödel's Way

Exploits into an undecidable world

Gregory Chaitin,
Newton da Costa &
Francisco Antonio Doria

CRC Press
Taylor & Francis Group
Boca Raton   London   New York   Leiden

CRC Press is an imprint of the
Taylor & Francis Group, an **informa** business

A BALKEMA BOOK

CRC Press/Balkema is an imprint of the Taylor & Francis Group, an informa business

© 2012 Taylor & Francis Group, London, UK

Typeset by MPS Limited, a Macmillan Company, Chennai, India
Printed and bound in CPI Group (UK) Ltd, Croydon, CR0 4YY

Library of Congress Cataloging-in-Publication Data

Applied for

Published by: CRC Press/Balkema
                 P.O. Box 447, 2300 AK Leiden, The Netherlands
                 e-mail: Pub.NL@taylorandfrancis.com
                 www.crcpress.com – www.taylorandfrancis.co.uk – www.balkema.nl

ISBN: 978-0-415-69085-0 (Pbk)
ISBN: 978-0-203-16957-5 (eBook)

On ne reçoit pas la sagesse,
il faut la découvrir soi-même,
après un trajet que personne
ne peut faire pour nous,
ne peut nous épargner,
car elle est
un point de vue sur les choses

Marcel Proust

Contents

Prologue

H OW DID YOU GET *your idea about the complexity of sequences? asks Doria. Chaitin answers: No, I'm not going to tell you, it's a long story... Well, ok, I'm going to answer you. When I was fifteen...Doria then interrupts Chaitin: When you were fifteen? When I was fifteen I was* sunbathing *on the beach, chasing girls...*
— *Oh, but you live in Rio!*

It was May 1994, and Greg Chaitin and Chico Doria were having a snack during an interval at the workshop on "Limits of Science" organized by John Casti and Joe Traub at the Santa Fe Institute. Newton da Costa had been invited to attend, but suggested that Doria be invited instead.

More than a decade later, in August 2007, the three authors meet again at a Rio workshop "Einstein and Gödel," which was sponsored by the Brazilian Research Center on Physics and the Brazilian Academy of Philosophy. The workshop's idea was to try to mingle Einstein's and Gödel's contribution. We may say that this book began to take form at that meeting and got its final form in a discussion at a table in an open air bar — *a* botequim, *so they are called* — *in Rio, Greg Chaitin, Virginia Chaitin, Chico Doria, and Newton da Costa consulted by phone, as he lives 600 miles away from Rio, in southern Brazil.*

In a nutshell, this book makes the case for the following claim:

Undecidability   and   incompleteness   are   everywhere   in   mathematics.

We could describe this book as a personal account by some of the participants in the work going beyond Gödel by finding uncomputability and incompleteness in many areas of continuous and discrete mathematics and theoretical physics. Still, our goal in this book is to show why it is everywhere, given our current conceptions about mathematics. And that not only we have to live with that as a basic fact of science, but it opens up new vistas and so far several quite enriching new ideas in the development of mathematics.

That is what we wish to assert in this book.

Gödel published his remarkable incompleteness theorems in 1931, and the imme-diate reaction is described in Ladrière's treatise, Les Limitations Internes des For-malismes. _Several efforts were made to show that there was a gap or a flaw in Gödel's argument, and even a book (published in 1933) which presents a brief sketch of Gödel's argument[1] exhibits some caution when the author says, "K. Gödel of Vienna seems to have proved ..." However it was soon noticed that we can derive an incompleteness result out of Church's undecidability theorem or out of Turing's unsolvability of the halting problem._

Then people began to ask whether incompleteness in mathematics would remain a far off nuisance which never interfered in the everyday work of professional mathemati-cians. Such was, for example, the belief (and hope) of René Thom. Nevertheless it was soon realized that undecidability and incompleteness appear everywhere in mathemat-ics, and not just in bordeline situations in arithmetic or set theory, as we know from results by S. Shelah (independence of Whitehead's conjecture from the axioms of set the-ory) or the celebrated Paris–Harrington theorem, which exhibits a perfectly reasonable example of an arithmetic statement which cannot be proved in Peano Arithmetic.

We can add a few landmark results in this pursuit. First, we have Cohen's 1963 proof of the independence of the continuum hypothesis and of the axiom of choice, from the axioms of Zermelo–Fraenkel set theory, supposed consistent. Then the Paris–Harrington theorem, which was published in 1979 as the final chapter of the Handbook of Symbolic Logic and which exhibits an arithmetical sentence with an obvious mathe-matical meaning and which is independente of Peano's axioms — again supposed con-sistent. The Paris–Harrington theorem sparked a series of results by Harvey Friedman in the same direction, where one exhibits formal sentences with mathematical relevance which are independent of several consistent, strong axiomatic systems.

However: does undecidability and incompleteness affect sciences which use math-ematics as its main tool? Yes, it does. Scarpellini's 1963 results can be interpreted as statements about electric circuits. Kreisel discussed the matter at length in an essay published in 1975; Richards and Pour-El considered several situations in physics; Benioff applied Cohen's forcing technique to physics. Outstanding research in that direction has been done by C. Calude and K. Svozil. We must also mention W. Myrvold's 1993 result on the undecidability of entangled quantum systems, a result that directly bears on the recent experiments about the Einstein–Podolsky–Rosen 1935 paper, on the tests of Bell's inequalities, quantum computation, teleportation, and so on.

Undecidability and incompleteness go even farther. Alain Lewis proved indepen-dence results in economics, and showed the undecidability of recursively presented Nash games. Again Vela Velupillai proved several undecidability results in economics. Cris Calude was kind enough to present the authors with a — nonexaustive, but extended — list of researchers that have contributed to Gödelian matters, so to say: J. Baez, J. Barrow, G. Boolos, S. R. Buss, J. Casti, M. Davis, A. Ehrenfeucht, H. Friedman, A. Kanamori, P. Lindström, G. Longo, Y. Manin, J. Paris and L. Harrington, S. Shapiro, S. Shelah, C. Smoryński, R. Solovay, M. Stay, G. Takeuti, D. E. Willard, W. H. Woodin, A. Yao, R. Zach.

---

[1]M. Black, The Nature of Mathematics, _Routledge and Kegan Paul (1933)_

This book discusses a piece of that action, certainly not the whole picture. Inspired by Chaitin's work, Newton da Costa and Chico Doria obtained several undecidability and incompleteness results in physics and elsewhere. They mainly stem from two basic results, the undecidability of chaos — is there an algorithm to test for chaos in dynamical systems theory? No, there are none — and the undecidability of equilibrium (stable or unstable?) again in dynamical systems. These results in turn originate in a very general undecidability theorem for classical analysis (that is, calculus) which mirrors Rice's theorem in computer science. We then show the relation of the da Costa–Doria results to Chaitin's work.

Chapter 2 and portions of the last chapter were written by Chaitin. Chico Doria drafted the rest of the book while exchanging views & comments with Chaitin, who was then in Rio. Newton da Costa contributed a thorough technical revision of the whole text.

We did not shy away from controversial or not fully completed results, but made explicit when it is the case. Anyway these more speculative matters were left for the two last chapters. Also this isn't a standard textbook; it is a kind of personal statement and as such we've added lots of "human interest" features and details. After all, mathematicians are flesh and blood creatures.

We hope that the readers will enjoy reading it.

Acknowledgments

G. J. Chaitin, N. C. A. da Costa, and F. A. Doria are members of the Brazilian Academy of Philosophy. This book was written while Chaitin was a visiting professor at the philosophy of science research program (HCTE) at the Federal University in Rio de Janeiro (UFRJ), March–May 2010. Newton da Costa and F. A. Doria acknowledge fellowships from CNPq, Philosophy Section (Ministry of Science and Technology, Brazil). We acknowledge support from the Production Engineering Program at COPPE/UFRJ and from its chairmen R. Bartholo, F. Zamberlan and S. Jurkiewicz. Also we thank HCTE and its director R. Kubrusly for providing financial and academic support to Chaitin.

Several friends contributed at different moments with suggestions and criticisms to the ideas presented in this book: E. Agazzi, J. A. de Barros, R. Bartholo, J.-Y. Béziau, E. Bir, A. Bovykin, O. Bueno, C. Calude, W. Carnielli, J. Casti, O. Chateaubriand, C. A. Cosenza, C. Doria, M. Doria, I. D'Ottaviano, S. French, S. Fuks, L. Gordeev, M. Gleiser, M. Guillaume, R. Koppl, D. Krause, R. Kubrusly, D. Miller, D. Mundici, J. R. Moderno, M. Novello, M. Paty, A. Pinto, A. Rodrigues, P. Suppes, V. Velupillai, S. Zambelli. We gratefully acknowledge their contribution.

Finally we would like to acknowledge the help — and patience — of Germaine Seijger, Jose van der Veer and Lukas Goosen, from CRC Press/ Balkema, during the preparation of this book.

About the Authors

Gregory Chaitin (1947) is an Argentinian-American mathematician and computer scientist. The author of many books and scholarly papers, Chaitin proved the Gödel-Chaitin incompleteness theorem and is the discoverer of the remarkable Omega number, which shows that God plays dice in pure mathematics. Currently, he is attempting to create a mathematical theory of evolution and biological creativity, based on considering life as evolving software. He is a member of the International Academy of the Philosophy of Science and of the Brazilian Academy of Philosophy, and was awarded honorary doctorates from the University of Cordoba and the University of Maine. Chaitin is currently a CAPES visiting professor at the Federal University of Rio de Janeiro (UFRJ) in the program on Epistemology and History of Science and Technology (HCTE). He is also an honorary professor at the University of Buenos Aires.

Newton da Costa (1929) is a Brazilian logician who is best known contribution has been in the realm of nonclassical logics. Da Costa developed paraconsistent logics, that is, logical systems that admit inner contradictions. Da Costa has wide-ranging interests, which go from foundational issues in the philosophy of science to physics (general relativity and quantum theory); besides his development of paraconsistent logics, he introduced the concept of quasi-truth to deal with mutually inconsistent scientific theories. Da Costa has a B. Sc. in civil engineering and a PhD in mathematics. He has visited several major universities (Stanford, Berkeley, Paris VII among others) and published about 200 scientific papers and several books on logic and the foundations of science. In 2009, he became a Professor Emeritus at Unicamp (Campinas, Brazil). Newton da Costa is a member of the Institut International de Philosophie, of the International Academy of the Philosophy of Science and of the Brazilian Academy of Philosophy.

Francisco Antonio Doria (1945) is Brazilian physicist. Doria is a Professor Emeritus at the Federal University of Rio de Janeiro, where he currently teaches economic theory at the graduate School of Engineering (UFRJ COPPE). Doria has a B. Sc. in chemical engineering and a PhD in mathematical physics. He has made contributions to the gauge field copy problem in quantum field theory and proved with Newton da Costa several incompleteness theorems in mathematics, physics and mathematical economics, including the undecidability of chaos theory. Doria is a member of the Brazilian Academy of Philosophy, was a Senior Fulbright Scholar at Stanford University, 1989–1990, and a visiting researcher at the mathematics department, University of Rochester.

A Caveat

To write about science is to walk on a razor's edge. If one adds too much technical detail one loses readability; if we wave our hands too much, precision and correctness may be lost. The authors are well aware of the fine points in the discussion, e.g. they know that one must suppose that our formal systems are consistent, that they must contain formalized arithmetic in a very precise way in order to be able to prove the Gödel incompleteness results, and so on. When they talk about mechanical procedures, they have Turing machines or equivalent formulations in mind; when they say that a consistent formal system cannot prove its own consistency, they mean the formalized sentence that Gödel exhibits, and which can be interpreted as the assertion of the system's consistency. Also, glitches and faux pas are unavoidable, but we tried to minimize them.

There are many fine points in our exposition. We try to refer to them but in order to ensure fluency and a readable text we do not hesitate to wave our hands a lot and to sacrifice rigor in order to produce a text that can be understood by a larger audience. The bibliography we exhibit tries to be exhaustive and contains papers and books with all details we've skimmed over; it also contains items we think are especially relevant to the topics in the present book. The interested reader is invited to browse them.

1. Gödel, Turing

C AN WE COMPUTE EVERYTHING? Can we solve all kinds of mathematical problems with some mechanical device? Can we build a
kind of mechanical theory that proves all mathematical truths?

No. We can't. That's what undecidability is about.

From 1931 to 1936 several mathematicians published a flurry of papers
where they discuss (and answer in the negative) the questions we have just
asked. In 1931 Kurt Gödel, an Austrian mathematician, published a paper
where he exhibits an obviously true sentence that cannot be proved or disproved in several versions of formalized (axiomatized) arithmetic — more precisely, a sentence so that neither it nor its negation can be proved in the usual
axiomatic frameworks for arithmetic. Follows papers by Alonzo Church,
Stephen Cole Kleene, and finally, one by Alan Turing, presented in November
1936 and published in 1937. Turing, a British mathematician who was still pur-
suing his PhD (he will obtain it in 1938 with another remarkable contribution)
defines our current concept of mechanical computation.

Problems that cannot be mechanically solved are *undecidable problems*.
Formal sentences that can neither be proved nor disproved within the cadre
of some reasonable, accepted mathematical theory, exhibit the phenomenon
of *incompleteness*. Both are related: as we will show by going through some
ideas by the Polish American logician Emil Post, undecidability in mechanical
devices leads to incompleteness in formal theories. Also, they can be super-
seded — we can get theories that prove many more arithmetic truths if we
add to them the so-called nonconstructive, or infinitary, rules. And there are
hopes that undecidability will be conquered, if we may say so, by hypercomputation theory, which so far is untested. (The sentences which can neither be
proved nor disproved within one such axiomatic theory are called *undecidable
sentences*, stressing the relation between undecidability and incompleteness.)

A consequence of Gödel's incompleteness theorems is the fact that if a
formal mathematical theory with enough arithmetic is consistent, that is, if it
doesn't contain any contradictions, then it cannot prove its own consistency
out of its axiomatic framework. Namely, the theory includes a mathematical
formula which effectively expresses the theory's consistency, and yet that formula cannot be proved within the theory.[2]

Does that mean that we will never be free of the fear that a contradiction
will come out of, say, arithmetic or even larger portions of mathematics? No,
for there are reasonable arguments that allow us to prove the consistency of

---

[2]We will discuss the meaning of "effective" later on.

those theories. Yet, again, they are nonconstructive, but reasonable and quite intuitive, as we will show.

Finally, do the undecidability and incompleteness results affect mathematicians in their everyday work? Do they come up in ordinary, run-of-the-mill mathematics?

We claim that the answer is yes! They do.

We will see that undecidability and incompleteness are everywhere, from mathematics to computer science, to physics, to mathematically-formulated portions of chemistry, biology, ecology, and economics. The Gödel–Turing phenomenon is an all-pervasive fact that has to be understood in order to be conquered and usefully mastered.

Gödel: logic and time

It is easy to find on the web the picture we are going to describe; just connect two names — Einstein, Gödel — and google them. Select "images," and, there it is, in full color. To the left you have a familiar figure, Einstein. We see an old gentleman with white dishevelled hair, a thick white moustache, a blue crumpled sweatshirt and what looks like a pen nonchalantly pinned to the sweatshirt's collar. To the right you see his much younger companion, Gödel, in a strict formal double-breasted suit, a carefully poised tie under the rigid collar, in full contrast to his companion. The younger gentleman has thick black-rimmed round glasses, his hair is brownish without any strands of white color, and is carefully, very precisely combed — one notices the straight dividing line that goes from his forehead to the back of his head.

Both are smiling; they look good friends and are in fact close friends. The old gentleman is, again, Albert Einstein, the physicist. The youngish-looking gentleman is Kurt Gödel, the logician.

That photograph is dated sometime around 1950. Gödel was then in his forties and Einstein was around seventy years of age.

Some twenty years later we have another vignette about Gödel:

> *His thought content was somewhat paranoid, and he had fixed ideas regarding his illness, and my diagnostic impressions were that there was indeed a severe personality disturbance, with secondary malnutrition, and some somatic delusions — inappropriate ideas about his bodily structure and function.*

This is a psychiatric assessment made by a physician established at Princeton [73], Dr. Harvey Rothberg, of Gödel's mental condition. Gödel had a long history of physical and psychological ailments, and he had already been treated in Viennese sanatoria in 1934 and 1936.

There is a third personal vignette about Gödel that we would like to quote before we enter our main subject:

> *He did, however, have in his youth the strength to pursue women. "There is no doubt," wrote a college friend, Olga Taussky–Todd, "about the fact*

that Gödel had a liking for members of the opposite sex, and he made no secret about this fact." Gödel, she went on, was not beyond showing off his acquaintance with a pretty face. Taussky–Todd herself, to her dismay, was once enlisted to come to the mathematical aid of one such young woman who in turn was trying to make an impression on Gödel. Was this interest in women confined to Gödel's youth? Not if his wife, Adele, is to be believed. Teasing her husband, she quipped that the Institute for Advanced Study [...] was packed with pretty female students who lined up outside the office doors of the great professors.

([156], p. 10). We may add, very much like a scene out of an Indiana Jones movie, with a bunch of girls waiting outside Indiana's office...

Do these vignettes help us in describing the complex personality of Kurt Gödel? Do they elucidate his strange creativity?

Well, three more anedoctes before we start doing business.[3]

I should also tell you about the fact that Gödel liked to go to night clubs, nothing unusual for well-to-do young men in Vienna. What was unusual was that he also did top quality math. He met his wife in a night club. She was not a dancer, which more or less would imply she was a prostitute. She was the hat check girl at the coat closet. A friend of mine Jimmy Schimanovich who knows a lot about Gödel tried to take me to this night club, but they wouldn't let us in. Then we noticed it was now a gay club; maybe we looked too hetero. It was in a Vienna basement; I vaguely remember walking down the stairs with Jimmy.

The Martin Davis story is that when he was a visitor at the Princeton IAS, he used to drive to work. One day he saw a strange, big bag lady (dirty person who lives on the street) plus a small well-dressed lawyer walking down the middle of the road with their backs to the traffic (not crossing the street, using it) lost in conversation. Can you guess who they were? For one mad moment he was tempted to run them over and go down in history as the crazy person who killed the greatest physicist and the greatest mathematician of the century! (That is, Einstein and Gödel.)

The editor of Scientific American *story is as follows: he (Dennis Flanagan) was living in Princeton, and had just published the 1956* Scientific American *article "Gödel's Proof" by Nagel and Newman, later a small book. So he knew how Gödel looked, because they had photographed him for the article. One hot summer day, he was walking in Princeton, and saw Gödel approaching. He decided to introduce himself. However Gödel stopped dead in his tracks to admire a sexy coed (female college student) who wasn't wearing very much because of the heat + humidity, and la belle opportunité s'est perdue. Dennis Flanagan didn't dare to disturb Gödel, who was obviously concentrating on the young beauty (something which is not considered proper form in the US, but perhaps is not so unusual in Vienna, a rather hedonistic city).*

---

[3]The narrator is one of the authors, GJC.

(And also not unusual in Rio, another hedonistic city, we should add at this point.)

A short biography

Kurt Gödel was born in April 28, 1906, in Brno, Moravia, of Lutheran parents. He died at Princeton, NJ, in January 14, 1978, of self-imposed starvation as he believed his food was being poisoned and therefore refused to eat anything.

He studied mathematics and physics at the University of Vienna and passed his PhD in 1930, when he was 23. His thesis was his first major result, as he proved the completeness of the predicate calculus in it. Then comes his great breakthrough: in a paper called "On formally undecidable sentences of *Principia Mathematica* and related systems," published in 1931,[4] he obtains his famous incompleteness theorems: for an axiomatic system with enough arithmetic in it (we will later clarify that) and whose theorems can be listed by some mechanical procedure,

- If that system is consistent, then there is a sentence in it which is true, but which can neither be proved nor disproved in the system.

- If the system is consistent, then a formal sentence that asserts its consistency cannot be proved within the system itself.

We'll later add more details to those ideas. Anyway, the thing is quite astounding: for suppose that you write down the undecidable sentence and add it as a new axiom to the system's axioms. Then the extended system will have a new undecidable sentence, which is by the way also undecidable in the original system. So there is an infinite listing of such undecidable sentences, given any of the usual axiom systems one uses in mathematics such as axiomatic set theory or formalized arithmetic.

Gödel becomes famous, is invited to the United States where he gives talks at the Institute for Advanced Study at Princeton, suffers a nervous breakdown back in Austria in 1935; is forced to emmigrate to the US in 1940 and settles down at Princeton NJ for the rest of his life.

At Princeton he becomes a close friend of Einstein's, for whom he writes two papers on general relativity (with some very original features, as the so-called Gödel universe exhibits a kind of natural time machine in it). He also proves the consistency of the continuum hypothesis and of the axiom of choice with the axioms of set theory at that same time. He had married Adele Nimbursky, formerly Porkert, in 1938 after his father's death, since his father had been opposed to their relationship as Adele worked at a Viennese cabaret and was six years older than Gödel.

From the 1950s on he becomes more and more interested in philosophical questions, and his only recently published notes from that period range from

---

[4]K. Gödel, "Über formal unentscheidbare Sätze der Principia Mathematica und verwandter Systeme," *Monatshefte für Mathematik und Physik* **38**, 173–198 (1931)'

a formal treatment of Anselm's proof of the existence of God — Gödel studies in depth its logical structure — to investigations on Husserl's phenomenology.

His previous nervous breakdowns develop into near paranoia and he dies of starvation in early 1978.

Some more final vignettes, from Rebecca Goldstein's book (see the references):

> Though Princeton's population is well accustomed to eccentricity, trained not to look askance at rumpled specimens staring vacantly (or seemingly vacantly) off into space-time, Kurt Gödel struck almost everyone as seriously strange, presenting a formidable challenge to conversational exchange. A reticent person, Gödel, when he did speak, was more than likely to say something to which no possible response seemed forthcoming:

> John Bahcall was a promising young astrophysicist when he was introduced to Gödel at a small Institute dinner. He identified himself as a physicist, to which Gödel's curt response was "I don't believe in natural science."

> The philosopher Thomas Nagel recalled also being seated next to Gödel at a small gathering for dinner at the Institute and discussing the mind-body problem with him, a philosophical chestnut that both men had tried to crack. Nagel pointed out to Gödel that Gödel's extreme dualist view (according to which souls and bodies have quite separate existences, linking up with one another at birth to conjoin in a sort of partnership that is severed upon death) seems hard to reconcile with the theory of evolution. Gödel professed himself a nonbeliever in evolution and topped this off by pointing out, as if this were additional corroboration for his own rejection of Darwinism: "You know Stalin didn't believe in evolution either, and he was a very intelligent man."

> "After that," Nagel told me with a small laugh, "I just gave up."

> The linguist Noam Chomsky, too, reported being stopped dead in his linguistic tracks by the logician. Chomsky asked him what he was currently working on, and received an answer that probably nobody since the seventeenth-century's Leibniz had given: "I am trying to prove that the laws of nature are a priori."

> Three magnificent minds, as at home in the world of pure ideas as anyone on this planet, yet they (and there are more) reported hitting an insurmountable impasse in discussing ideas with Gödel.

This is the man. We are now going to take a glimpse at his work.

The incompleteness theorems, I

Gödel's original argument makes a brief reference to the well-known Epimenides Paradox, "Epimenides the Cretan says: all men are liars." Gödel builds a formal sentence noted G in some axiomatized version of arithmetic; that sentence is self-referential and translates as:

> G *cannot be proved.*

(We mean: it cannot be proved in the given axiomatic framework.) Then suppose that our system is consistent, and moreover suppose that G is proved. It follows that G cannot be proved, by the very definition of G. A simple argument based on the fact that our system isn't supposed to prove false assertions shows that not-G also cannot be derived, if we are within a consistent system.

The whole argument is impeccable and flawless, but the weirdness of the unprovable sentence exhibited by Gödel raised doubts whether that kind of phenomenon might affect more substantial mathematical statements. And yes — it does.

Kleene's version of the first incompleteness theorem

Let's take a close look at Kleene's version of the first incompleteness theorem, which has a very simple argument behind it. Then we will come back to more, let us say, traditional presentations of the result.

Here we have Kleene's argument in a nutshell:

- For our purposes here, a *total computable function* is a function from the natural numbers $0, 1, 2, 3, \ldots$ with integer values which is bug-free, that is, given any number n, we feed it into the program that computes our function, and obtain some numerical value as its output.

- Suppose that our axiomatic theory is consistent and can "talk" about arithmetic; at least about such total computable functions.

  Also suppose that there is a computer program that can list all theorems in our theory, which we designate by S. (We can do that for axiomatic set theory or for formal arithmetic, even if the proofs may be very long and cumbersome.)

- Start the listing of the theorems.

- Pick up those that say: "function f is total computable."

- Out of that we can build a list $f_0, f_1, f_2, \ldots$, of total computable functions in S with their values like this:

$$f_0(0), f_0(1), f_0(2), f_0(3), \ldots$$
$$f_1(0), f_1(1), f_1(2), f_1(3), \ldots$$
$$f_2(0), f_2(1), f_2(2), f_2(3), \ldots$$
$$f_3(0), f_3(1), f_3(2), f_3(3), \ldots$$
$$\ldots$$

- Now define a function F:

$$F(0) = f_0(0) + 1.$$
$$F(1) = f_1(1) + 1.$$
$$F(2) = f_2(2) + 1.$$
$$\cdots$$

- F is different from f_0 at value 0, from f_1 at 1, from f_2 at 2, and so on.

We can now conclude our reasoning. The $f_0, f_1, f_2, \ldots$ functions are said to be *provably* total in our theory S, as they appear in the listing of the theory's theorems. However F cannot be provably total, since it differs at least once from the functions we have listed. Yet F is obviously computable, and given programs for the computation of $f_0, f_1, f_2, \ldots$ we can compute F too.

So the sentence "F is total" cannot be proved in our theory.

Also, if we suppose that the theory is *sound*, that is, if it doesn't prove false facts, then the sentence "F isn't total" cannot be proved too. Therefore "F is total" is an undecidable sentence within our theory.

This is Kleene's 1936 argument. We will later on present the more extensive but illuminating Gödel–Turing–Post argument.

An immediate consequence of Kleene's proof

One of the most celebrated unsolved mysteries of current day mathematics is the P vs. NP problem. Briefly, it deals with problems which are such that, if we have a solution for one of its instances, it is easy (fast) to check it for correctness, but all known procedures require a long time and much computer effort to arrive at one solution, in the general case.

For "easy" or "fast" one should understand, "in polynomial time." For "hard," read "in exponential time." With some more detail: a computer program that operates within the bounds of polynomial time is such that its operation time is limited by a polynomial function on the length of its input measured in bits. There are several examples of such problems, which constitute the NP class, such as the traveling salesman problem or assigning classes, students and teachers to a restricted number of classrooms and class hours.

The $P = NP$ conjecture may be formulated as a question:

Is there a fast program that settles all problems in the NP class?

In order to solve it we must consider the set of all fast programs, and somehow try our instances of NP problems to each one of them. But can we precisely know which are the fast algorithms?

Consider a very strong axiomatic theory, such as set theory plus several powerful large cardinal axioms. Then the following sentence describes an assertion which is unprovable in that theory, as well as its negation:

There is a set of programs so that even such a powerful theory cannot decide whether it is a set of fast programs or not.

This is quite remarkable, as it affects a practical situation. Its proof stems out of Kleene's incompleteness result.

Actually we can go beyond that: we can exhibit formal sentences in arithmetic such that the sentence "P is a polynomial Turing machine" is undecidable, that is P is polynomial in one interpretation for the theory, and exponential in a different one. Formalized computer science is riddled with undecidable stuff.

The incompleteness theorems II: consistency cannot be proved within the system

A formal system is consistent if, whenever one proves some sentence A in it, no proof of the negation of A can be found. Let's make it more precise: a formal system S is consistent if there is no sentence A in it so that neither A nor its negation not-A (or $\neg A$) can be proved in S.

Why is it important for a system to be consistent? Because once you have obtained one contradiction in it, by the rules of classical logic one can prove everything in the system. So the system is trivialized, that is every sentence which can be written in its language can be proved.

If the system includes enough arithmetic, given inconsistency one then proves absurd statements such as $0 \neq 0$, or $0 = 1$, and so on:

> *If the system is based on classical logic and cannot prove at least one sentence of its language, it is consistent.*

One then formalizes the idea of consistency as the following statement:

> *A formal system which includes arithmetic is consistent if and only if one cannot find a proof of $0 \neq 0$ in it.*

Again call the system we are considering S. Then we can abbreviate the consistency statement as $\mathrm{Con}(S)$. Gödel proves in his 1931 paper that neither $\mathrm{Con}(S)$ nor its negation can be proved in S, if S itself is consistent. So, both sentences $\mathrm{Con}(S)$ and not-$\mathrm{Con}(S)$ cannot be proved in S.

The proof goes as follows: one shows that the following conditional sentence can actually be proved in our formal system:

> *If* $\mathrm{Con}(S)$ *holds then so does G.*
>
> *Formally:* $\mathrm{Con}(S) \rightarrow G$.

(G is Gödel's original sentence, which asserts its own unprovability.) There is an intuitive explanation for the implication above: it means that if system S is consistent, then at least one formal sentence will not be proved in S — and G asserts its own unprovability. In fact G can be seen as a definition of $\mathrm{Con}(S)$, as G asserts its own unprovability.

Then, one should recall the following classical well-known logical argument:

If A then B.

Therefore if not B, then not A.

Think of an easy example: *if there is thunder then there will be rain.* By contra-
position, *if there won't be rain then there is no thunder.*

The argument for the second incompleteness theorem can now be completed out of the preceding comments.

- We know that S proves: If Con(S) holds then so does G.

- Then suppose that S proves Con(S).

- If S proves Con(S) then it proves Gödel's sentence G.

- However S doesn't prove G.

- Then S cannot prove Con(S).

The whole argument hinges on the sentence: If Con(S) holds then so does
G. As we've pointed out, it's a conditional sentence which can be proved within
formalized arithmetic.

This is the content of Gödel's second incompleteness theorem.

Now we ask, can we substitute G for Kleene's sentence "F is total" here?
No. In fact we have that Kleene's sentence leads to a different conditional
statement, which can be proved (for different F, each one adequate to the corresponding axiomatic environment where they are cradled):

If F is total recursive then we have that Con(S) *holds.*

Sentences G and "F is total" have different structures; G is an example of what
is called a Π_1 sentence, as well as Con(S), while "F is total" is a Π_2 sentence.[5]

A weird formal system

Now add Con(S) to system S. We obtain a new consistent system which proves
the consistency of S. However what is the meaning of system S together with
sentence not-Con(S)? Such a system is also consistent if S is consistent — but
it proves a sentence that asserts that S is inconsistent!

Let's take a closer look at this apparent paradox.

Notice the form of the consistency statement which we now repeat:

A formal system which includes arithmetic is consistent if and only if one
cannot find a proof of $0 \neq 0$ in it.

---

[5] A Π_1 sentence is of the form "for all x we have that x has property P," while a Π_2 sentence
is, "fir all x there is an y so that x and y are related by relation R." P here should be decided by
computational means.

The catch is in the way we code the proof of $0 \neq 0$ or any proof in our formal system. It is coded by a number — in a way we will soon describe.

Anything that can be written with some alphabet can be given a numerical code. Just list all possible sequences of letters, plus a blank space, quotation marks, parentheses, punctuation signs etc, in order of length and then as in a dictionary. Do it for sequences of length zero — which we add for the sake of completeness — length one, two, and so on. Most sequences will turn out to be garbage, but we can select the meaningful ones and build up a new listing of those meaningful sequences, and give them a rank number. This is the simplest kind of *Gödel numbering*, a procedure originally devised by Gödel to make formal language sentences into arithmetical sequences.

Now: *the number that codes the proof of $0 \neq 0$ exists, but we cannot "open it up" because it is a nonstandard number.* Let us explain it.

We can extend the natural numbers in such a way that our mystery code number appears — but, lo!, it's a "nonstandard natural number," an immensely large, nondescript, but still finite natural number, which unfortunately we cannot decode to uncover the way we can do a proof of $0 \neq 0$ in our formal system.

It follows that while we can prove not-Con(S) in our theory, we have no procedure in S to find out how to get a contradiction in our theory.

Can we prove the consistency of arithmetic?

If you try to prove the consistency of some formal system for arithmetic, and if you ask for what mathematicians call a strict finitistic proof, that can only be done outside arithmetic. For instance, embed formal arithmetic within a stronger system such as set theory, and you can prove the consistency of arithmetic within that larger background. However we must then believe that set theory is consistent. Well, we can prove the consistency of set theory out of a still stronger theory, e.g. set theory plus one inaccessible cardinal (a very large infinite number that cannot be reached from smaller sets with the help of the tools and operations available within set theory).

Yet we must then prove the consistency of that bigger theory, and so on.

The first consistency proof for arithmetic was obtained by Gerhard Gentzen in 1936. Gentzen shows that we cannot simultaneously derive A and not-A in arithmetic.

Gentzen's proof demands quite some technical expertise, and isn't in fact totally transparent, as the main tool used is the rather abstract concept of "transfinite induction up to the constructive ordinal number ϵ_0." Gentzen's proof sort of exhibits all possible proofs in formalized arithmetic, and shows that $0 \neq 0$ cannot be derived as a theorem of the theory. But that's the interpretation of Gödel's consistency statement Con(S) within arithmetic.

However we can obtain a rather simplified, more transparent version of Gentzen's proof with the help of Kleene's F function, which for arithmetic reads as F_{ϵ_0}. We'll explain its meaning — essentially it is an outrageously complicated generalization of $+$ and $\times$.

Let's build a sequence of functions:

- F_1 is the sum $+$.

- F_2 is the product $\times$ (iteration of the sum).

- F_3 is the exponential (iteration of the product).

- ....

- F_n is the iteration of F_{n-1}.

- ...

- F_ω is the first operation with an infinite index (ω is technically the ordinal number that describes the way all natural numbers $0, 1, 2, 3, \ldots$ are ordered). It is obtained out of all the preceding functions and is called Ackermann's function.

- ...

- F_{ϵ_0}, where ϵ_0 is an ordinal defined as $\omega^{\omega^{\omega^{\cdots}}}$

The construction of F_{ϵ_0} may seem quite farfetched, but it is in fact a function that can explicitly be given a program; it can also be naïvely seen to be bugless, that is to say, it is a total computable function. It is also Kleene's F in the case of arithmetic (or at least a function with the same relevant properties as F), so we cannot prove it to be total in arithmetic.

We can prove within formalized arithmetic the sentence:

If F_{ϵ_0} is total then $\mathrm{Con}(PA)$.

Here PA stands for Peano Arithmetic, which is the usual axiomatization we take for arithmetic systems.

Now it is — naïvely always — clear by construction that F_{ϵ_0} is total. Therefore one gets that $\mathrm{Con}(PA)$. Kenneth Kunen exhibited in 1995 an algorithm that does precisely that: it proves the consistency of arithmetic. That algorithm of course leads to a function that cannot be proved to be total within PA.

As a final remark: we cannot prove Gödel's sentence $\mathrm{Con}(PA)$ within PA, if that theory is in fact consistent, but there are weaker, nonconstructive sentences which translate as consistency statements that can be proved within the theory whose consistency they purport to describe. That fact was explained in depth by Solomon Feferman in 1960.

Chaitin's incompleteness theorem

A more recent development which we are going to discuss at length later on is Chaitin's 1974 incompleteness theorem. Before we briefly comment on it let us notice that a major step in the argument was conceived by Chaitin in early 1971 in Rio, a few days before Carnival — Rio's environment seems to be quite fertile for good math, as about ten years earlier Steve Smale conceived his idea of a horseshoe attractor (a kind of strange attractor) while lying on the sands of Copacabana beach.

Chaitin had conceived in 1965 an idea that led to a definition of randomness in strings of bits: a string is random whenever it cannot be generated out of an universal computer by a shorter string.[6] So, random strings have programs that are about as long as the strings themselves. The smallest program that generates one string is its information content.

Chaitin's incompleteness theorem asserts that given some formal theory like our S, it cannot prove sentences like:

The information content of string x is larger than k, where k is an integer,

beyond a certain value for k, which depends on the axioms of S seen as a program for the enumeration of the theorems of S, and which is also related to the program that computes Kleene's function F for system S. So, everything begins to fit into place.

Moreover F begins to show here a monster-like face, as it is a kind of computable version of a noncomputable, fantastically fast-growing function known as the Busy Beaver function.

(Well, the goal of this paragraph is to act as a kind of teaser or sneak preview of things that will still come in this book.)

Berry's Paradox

Chaitin tells us that his incompleteness theorem was motivated by Berry's Paradox. Here goes the paradox: consider

The smallest positive integer not definable in under eleven words.

But this sentence has ten words, and purports to define it!

Godfrey George Berry (1867–1928) was a rather peculiar character. Here are a few reminiscences of his by one of his grandnephews, a university professor, as communicated to Chaitin:

> *Godfrey George came to quite a sad end…while at Oxford he deserted his wife and two daughters for what our family called "a barmaid." I met both daughters later in their lives (I was born in 1940). One daughter lived in Oxford all her life and never married – I met her when I was an undergraduate there in the late 1950s; the other daughter married a Scotsman. They had a millinery business in London and sold their wares to Queen Elizabeth (mother of the present Queen) and similar types of people. They eventually retired, and their children still have a flourishing mushroom business just north of Edinburgh.*

> *Well, to get back to George Godfrey: his sad end was when Phyllis, the daughter who lived in London was called to a hospital (or police station) to*

---

[6] But for a constant term.

help identify her father who was essentially comatose and had been found lying drunk in a gutter — I believe he died in a hospital a few days later. The family had been more or less out of touch since his departure which I believe was just before the First World War; probably 1914 or so.

My father always knew he was at the Bodleian — although sometimes the story was that he was a Balliol College, which he may well have also been — but his claim to fame in the family was that he could speak 14 different languages. He did the authoritative translation from the Greek of Thucydides, since supplanted by several more modern versions. And he also was the translator for several Danish texts on the education of young children.

So together with this connection, he was clearly a clever fellow, probably taking on lots of different projects during boring times as a Bodleian librarian...[7]

Rice's theorem

We now come to what can be seen as the most important consequence of undecidability in practical situations; or perhaps the most destructive consequence of the incompleteness phenomenon discovered by Gödel. Have you ever wondered why one cannot have an antivirus in our computers that doesn't have to suffer constant updates? Why is it that computer programs have so many bugs all the time and have to be constantly upgraded with patch–up subroutines?

The reason lies in Rice's theorem.

Henry Gordon Rice was born in 1920 and passed his PhD thesis in 1951 at Syracuse University; his famous result appear in that thesis but was only published in 1953. Basically it says the following:

Suppose that P is some nontrivial property of a class of computer programs, that is, there are programs which satisfy P and also those that do not satisfy P.

Then no computable procedure can distinguish among all programs those that hold for P from those that do not hold for P.

Think of P as, say, program x is a virus. Follows from Rice's theorem that one cannot have an universal vaccin for computer viruses (and if the DNA and RNA are seen as biological versions of computer programs, this applies to biological viruses as well).

Also we cannot test an arbitrary program for, say, specific bugs and the like. Therefore Rice's theorem makes computer science into a nearly empirical, trial and error endeavor.

We can derive Rice's theorem out of Gödel's incompleteness theorem. The argument is quite simple. Suppose our theory S with arithmetic in it, and suppose given a property P of some objects x in it. Moreover suppose that object a

---

[7]Personal communication to G. J. Chaitin.

satisfies P, while b doesnt't. Then consider the object described in the following sentence:

The object x such that: either $x = a$ or $x = b$.

Therefore x is either a or b.

Now let's add a catch to the sentence:

The object x such that: either $x = a$ and $\mathrm{Con}(S)$ — or — $x = b$ and not–$\mathrm{Con}(S)$.

While this looks contrived (and is in fact contrived), one can easily show that such an object is undecidable for P in theory S. For, say, if we choose the first alternative, it simultaneous decides that $x = a$ together with $\mathrm{Con}(S)$, which violates Gödel's second incompleteness theorem.

Now there cannot be an algorithm for deciding P in everyday computer practice. For we would then internalize it in S and decide our undecidable sentence.

(The above paragraph is, essentially, Rice's theorem.)

We conclude here our first, brief, overview of Gödel incompleteness in respect to everyday practice in mathematics. Does it matter? Yes, sir, yes, ma'am, it definitely does.

Gödel incompleteness deeply affects mathematics in its theoretical side and in its applications. From computers in our everyday life to very deep and abstract issues in the work of pure mathematicians.

More work by Gödel: the constructive universe of sets

The next great achievement of Gödel's after the two incompleteness theorems is his proof of the consistency of the continuum hypothesis and of the axiom of choice with the axioms of set theory (the so-called Zermelo–Fraenkel axiom system).

Set theory was created by Georg Cantor (1845–1918), a mathematician who was born at St Petersburg of German stock. The whole theory arose out of Cantor's efforts to describe the functions that can be represented as Fourier series, but it soon became much bigger than its original motivation. Cantor showed that there are infinitely many orders of infinite collections (sets) of numbers: the natural numbers $0, 1, 2, 3, \ldots$ have as their cardinal number (the number of all natural numbers) the *countable infinity cardinal*, which Cantor noted $\aleph_0$. He then showed that rational numbers (those that can be written as fractions) also form a countable infinity, and can in fact be placed in one to one correspondence with the natural numbers, a fact which was counterintuitive for several reasons.

How about the real numbers, the number of points in a straight line segment? In a brilliant tour de force Cantor showed that their cardinal number was a much greater infinity than $\aleph_0$, and he showed that the cardinal number of all real numbers in such a segment can be represented as $2^{\aleph_0}$. Of course:

$$2^{\aleph_0} > \aleph_0. \tag{1}$$

Now the question arises: is there any infinite set with a cardinality $\aleph_1$ in between? That is, is there some $\aleph_1$ so that one has[8]

$$2^{\aleph_0} > \aleph_1 > \aleph_0? \tag{2}$$

This is *Cantor's Continuum Problem*. The associated Continuum Hypothesis, sometimes abbreviated as CH, is the assertion:

There are no infinities between $2^{\aleph_0}$ and $\aleph_0$.

Cantor's question looked so mysterious and perplexing — after all it dealt with different orders of infinity, something that no one would have dared to conceive before Cantor — that when David Hilbert presented his list of 23 open problems to the International Congress of Mathematicians in 1900, it stands as the first problem at the top of the list.[9]

There were also theological overtones to the other question tackled by Gödel, the Axiom of Choice. For one argues that it is implicitly used in several of the traditional arguments used to prove the existence of God — according to some authors that believe to be following in the footsteps of Gödel.[10] Example:

> *There are good things in the universe. Given any good thing, we can assuredly find another better thing, and so on. Given all such chains of good things in the universe which are ordered by increasing goodness, we can certainly top them all with the* Summum Bonum, *the Maximal Good, which is God.*[11]

The implicit supposition, that given partial ordered chains we have an element that tops them all is here derived from the Axiom of Choice (we note, abbreviated as AC), here stated in the form of Zorn's Lemma, a proposition which was found equivalent to the Axiom of Choice. The original formulation of the Axiom of Choice explains its name:

> *Let there be an arbitrary family of mutually disjoint nonempty sets. Then there is a set which has one and just one element of each set in that family.*

Think of each set in the given arbitrary family as a bag. Then pick up a single element from each bag, and collect them all together in a new set, the "choice set." Looks naïvely true? Well, for finite collections of sets it certainly holds; also for some countable infinite families of sets. But for an arbitrary family, does it have to hold always?

---

[8]$\aleph_1$ has a precise definition which we skip here.

[9]There were also theological overtones to it, as Cantor was accused of being a pantheist (or of trying to "prove" that the universe was the realm of infinitely many gods) as he sketched an infinite hierarchy of higher and higher infinities, without end.

[10]We don't.

[11]There are several references to such a tongue-in-cheek (or not) proof of the existence of God, e.g., Robert K. Meyer, "God exists," *Nous* **21**, 345–361 (1987). The idea goes as above: we argue as in Aquinas, and then use Zorn's Lemma to prove the existence of a maximal element. However we might also argue that there are many maximal elements, and so we conclude that polytheism holds!

The Axiom of Choice is an essential tool. It is required to prove that an arbitrary vector space has a basis, for example. If we weaken it then the kind of quantum mechanics that arises out of a mathematical framework without the full Axiom of Choice is quite different from our usual quantum mechanics. (And of course if you are interested in scholastic theology, several of the proofs of the existence of God require it.)

The Axiom of Choice first came into being in modern mathematics in the proof of the well-ordering theorem:

> *Every set can be well-ordered, that is, all its segments are linearly ordered and have a least element.*

Actually the well-ordering theorem is equivalent to the Axiom of Choice, that is, one implies the other. It was explicitly included in the axioms for set theory by its formulator, Ernst Zermelo, in 1908, and is kept in the current version of those axioms, the Zermelo–Fraenkel (after Abraham Fraenkel) set theory with the Axiom of Choice, or ZFC theory for short. Zorn's Lemma, named after Max Zorn who conceived it in 1935, is another of the disguises of the Axiom of Choice.

There was much debate since the early 20th century whether the Axiom of Choice was acceptable as a general, basic mathematical principle. The question remained open until the 1930s when Gödel introduced the constructive universe of sets and proved that if the ZF (Zermelo–Fraenkel axioms, without Choice) are consistent, then both the Continuum Hypothesis and the Axiom of Choice are consistent with it.

Gödel's new development, that is, the proof of the consistency of the Continuum Hypothesis and of the Axiom of Choice, only appeared in 1938, but it seems that he had been working on it for quite a few years. It is a major tour de force of mathematical expertise.

The constructive universe is an universe which is built like an infinite high rise. When you reach level of order α, everything that one can build at alpha is supported by something in the preceding floors, $\alpha - 1, \alpha - 2, \ldots$, and so on. Thus, given any set at level α, it arises only out of previously constructed sets, which sit in the previous levels of the constructive universe. It's a highly organized structure, with nothing loose in it, and its very structure in successive levels indexed by ordinal numbers $\alpha, \alpha + 1, \ldots$, shows why the Axiom of Choice (as the Well-Ordering Principle) holds in it.

The constructive set-theoretic universe is also a kind of very spare model for the axioms of set theory, and this in turn more or less explains why the Continuum Hypothesis holds in it: there aren't enough sets to aggregate as intermediary kinds of infinities between $\aleph_0$ and $2^{\aleph_0}$. Of course this is just a big waving-hands argument; Gödel's full developmnt of the constructive universe is a painstaking, extremely careful task, and the proof that it satisfies the ZF axioms, that is, that it is a model for Zermelo–Fraenkel set theory, is a most delicate affair. But when we clear up the mists we see that the constructive universe is a natural generalization of our finitistic ideas about computability when extended to the ordinal number system.

We will later sketch how Paul Cohen proved in 1963 that the Continuum Hypothesis and the Axiom of Choice are independent o the ZF axioms.

A concluding note: Gödel on time machines

In 1949 Gödel joined other researchers in the organization of a Festschrift in honor of Einstein's 70th birthday. Instead of submitting a paper on logic, or on the philosophy of mathematics, Gödel decided to discuss a very weird kind of solution to the Einstein gravitational equations. Let us read the opening lines of Gödel's paper:

> *All cosmological solutions with non-vanishing density of matter known at present have the common property that, in a certain sense, they contain an "absolute" time coordinate, owing to the fact that there exists a one-parametric system of three-spaces everywhere orthogonal on the world lines of matter. It is easily seen that the non-existence of such a system of three-spaces is equivalent with a rotation of matter relative to the compass of inertia. In this paper I am proposing a solution (with cosmological term $\neq 0$) which exhibits such a rotation.*

The idea stated in the preceding paragraph is: until then all explored solutions to the Einstein gravitational equations described an universe that smoothly evolves along time. If we do not have that kind of universal or absolute or cosmic time coordinate then we have some kind of intrinsic rotation in the universe. (This absolute time coordinate is what allows us to say that, according to the current cosmological views, that the universe has an age of 13.7 billion years; without a cosmic time coordinate, everywhere defined, such an assertion would be meaningless.)

So, Gödel's solution goes against the mainstream in cosmology. It wasn't however a first such solution: another weird solution to the Einstein gravitational equations had already been obtained by Cornelius Lanczos in 1924.

The Gödel universe has two important properties:

- It doesn't have a global time coordinate.

- It has "natural" time machines, that is time-like curves that go into the past.

Time-like curves are trajectories of everything that moves at sub-light speeds, that is, us, the planets, the galaxies, and so on. They are trajectories followed by objects in the physical world, but for light and zero-mass particles.

The Gödel solution is still some kind of an outcast when it comes to cosmological studies today. In 2007 the three authors met at the workshop *Gödel and Einstein: Logic and Time* in Rio.[12] The idea was to try to approximate physicists, logicians and philosophers of science around the contributions of Einstein and

---

[12]Chaired by Mario Novello and Doria; it took place at the Brazilian Center for Research in Physics (CBPF), August 2007.

Gödel. At that meeting Mario Novello asked the question, how frequent are Gödel-like solutions among all the solutions for the Einstein equations? Can we distinguish solutions with a global time coordinate from those without one such coordinate?

The answer to the second question is no:

> *There is no algorithm or computational procedure that allows us to decide, in the general case, whether a given solution to the Einstein equations has a global time coordinate.*

> *If we axiomatize general relativity within the framework of Zermelo–Fraenkel set theory, then there are infinitely many formal sentences like "solution g to the Einstein equations has the global time property," which are undecidable within the given axiomatic framework.*

So we can marry both Gödel contributions, incompleteness and his weird cosmological solution.

And how about the first question? How frequent are the Gödel-like solutions? Of course there are uncountably many Gödel-like solutions, as well as again uncountably many solutions with absolute time, for the Einstein equations. But mathematicians have other tools to characterize large or small sets, like for instance probability. Then we may ask, how probable are the Gödel-like solutions? The answer is:

> *For a wide class of probability measures, the probability of finding a Gödel like solution among all solutions to the Einstein equations is 100%.*

A probability measure is a way of defining probabilities on a set of objects. For infinite classes of objects there may be several different, nonequivalent ways of doing so.) Does that mean that global time solutions do not exist? Certainly not; only, they are not the typical solutions. The typical universe has a very weird time structure.

But there is more in store. For if one carries things out to its strict mathematical conclusion, the most probable universe may be *exotic* and *set-theoretically generic*. A truly wild beast.

For a long, contented period, mathematicians had believed that once you define a topology for some curved space (curved spaces are called "manifolds" by mathematicians) you fix in an unique way the means of doing physics on it. Let's elaborate: giving a topology to a set means that you can describe neighborhoods; you know the way of dealing with qualitative concepts like "near" and "far," or "large" and "small".[13] Once you have a topology, that is you know how to talk about nearness in your set, some extra structure gives you the concept of distance.

Mathematicians thought that these structures were enough to describe how things move over trajectories on manifolds, so that one can do physics on them.

---

[13]Large sets are collected in structures called "filters," and small sets in "ideals." The term "ideal" comes from its use in 19th century algebra, and has no descriptive intent; "filter" is more concrete, as large sets are sort of arranged in it more or less like layers in a filter.

But in 1956 John Milnor surprised the mathematical world by announcing that the 7-dimensional sphere (a ball in seven dimensions) had 28 different, nonequivalent, smooth structures, that is, 28 different ways of defining velocities and accelerations on it. So, one cannot go in a straightforward way from topology to physics, that is, from topology to what one calls a differentiable structure. Milnor received the Fields Medal in 1962 for his discovery.

Anyway it came as a surprise the discovery since the late 1970s that 4-dimensional manifolds also have myriads of differentiable structures on top of a single topological structure. This stems from results by several researchers, like Sir Michael Atiyah, Sam Donaldson, Michael Freedman, Clifford Taubes. Taubes proved that the four dimensional hyperplane had in fact uncountably many differential structures, that is, uncountably many ways of defining velocities and accelerations over it.

Their work generated a shower of Fields Medals.

Why does it matter to our discussion? Well, four is the dimension of Einstein's spacetimes, that is, of our universe as depicted by general relativity. So, if we make our probability calculations, we will notice that our universe should also have one of those exotic differential structures.

And to top the pudding with a big, sweet cherry, and advance a few ideas we'll later discuss: it will also be set-theoretically generic, in Cohen's sense.

A truly wild, untamable beast.

Alan Turing and his mathematical machines

The last decisive event in Alan Turing's life happened in early 1952 when he invited a teenager he had picked up in Manchester to spend the night at his place. Later the boy was found to have taken part in a burglary at Turing's home. Then naïvely as he was questioned by the constable, Turing told that he had made love to the boy. He was then prosecuted under the old British anti-sodomy statutes which were still in force in 1952 and condemned to undergo a hormonal treatment to lower his sexual impulse. The treatment deformed Turing's body and in despair he killed himself with an apple laced on cianide in 7 june 1954, a few days before his 42nd birthday.

Gordon Brown, then the British Prime Minister, apologized for the government's mishandling of Turing in late 2009.

Turing was born in London on June 23, 1912, of well to do parents. He is best known for his conception of mathematical machines that follow a program in order to calculate the values of what we now call computable functions; they are now called Turing machines. But from 1939 on he worked at Blechtley Park, a research facility owned by the British government whose main purpose was to break the German military war codes. His contribution was decisive in solving the mystery of the Nazi *Enigma* machine, which encoded the main German military exchanges.

Later in life and a few years before his death, Turing became interested in morphogenesis and the evolution of organisms, and was a pioneer in the use of reaction-diffusion equations in the study of biological systems.

More on that: the original diffusion equation describes the way heat disperses in some medium; in the early 1900s chemists and biologists added the so-called reaction terms so that the resulting equation could now describe how compounds interact and disperse in a reaction container. It was then suggested by Alfred Lotka that the same kind of mathematical equations could describe biological phenomena.

We are going to be interested here in two of Turing's achievements:

- The development of Turing machines and the discovery of the halting problem.

- The development of oracle Turing machines and the concept of progressions of theories.

What is a computation?

Turing's main achievement has been the depiction so to say of a device that can be used in a very natural way to formalize the concept of computation — which we have been using until now without giving it the much needed clarification. Let's do it now.

Suppose that you sit down on your desk, pick up a pencil, an eraser, some paper, and start filling up your income tax statement. You have the instructions next to you in a booklet you've just printed from the internet.

You start by collecting the data — the input data — that is the total amount of salaries you've received, plus extra gains, deductible expenses and the like. You mix up everything according to the rules in the IRS booklet. You use the blank paper for auxiliary and side calculations, and after lots of sums and multiplications you arrive at the value of the tax you have to pay.

Let's make it explicit:

- You start from a finite set of data, the input.

- You follow rigid, deterministic instructions, which are also coded as a finite set of words.

- You use as much "draft" space as you require for side calculations, but it is always a finite amount of memory space. Memory is finite, but unbounded.

- The computation takes a finite amount of time.

- The output, or result, is given as a finite set of words.

- There is the possibility that in some complicated situations the computation will never end.

Turing developed his mathematical machines to study the decision problem:

> *Given a natural number n and some set C of natural numbers, can we always computationally decide whether $x \in C$?*

The answer turned out to be: no.

Turing machines, I

It is known that Turing's interest in the so-called decision problem arose out of a talk given at Cambridge by the British topologist M. H. A. Newman. Turing's ideas on the matter seem to have crystallized around 1935, when he conceived his theoretical computing device which is now the most intuitive explanation one has for the concept of algorithm. His great paper "On computable numbers, with an application to the *Entscheidungsproblem,*" was submitted for publication early in 1936 and orally presented in November 12 of the same year. However the initial reaction to that great piece of work was quite subdued.

The 1936 paper describes what we now know as Turing machines, shows that there is an universal Turing machine, that is, a machine which can simulate any other Turing machine, and proves the unsolvability of the halting problem. We will now examine in detail all these three results.

Turing machines, II

Turing once compared his device to a very simple typewriter. Let's follow his analogy. Think of a printing device with some kind of control attached to it. The printing device's head runs over an infinite tape divided into squares. The printing head sits at each step over one single square on the tape. It reads whatever is written on the square — the square may be blank, or have on it a 0 or a 1 — and according to its "internal state," it may keep what is written on the square, or erase it and write a different sign (or leave it blank) and then move right, left, or even go to a "shutdown" state.

It's simple; let's recapitulate:

- We have an operating head and a tape divided into squares under it. The tape is potentially infinite, but during each computation we will only use a finite portion of it.

- The head is positioned over a square on the tape. The square is either blank or has a 0 or 1 on it.

- The head is in a given internal state.

- The head reads what is written on the tape's square under it.

- Propelled by its internal state, it acts: either leaves it as it is, erases the square, or writes a 0 or 1.

- Then it moves right or left.

- And the head's internal configuration goes to another state, which may be a "shutdown" state that ends the machine's operation.

(The shutdown state is the state on which the machine stops after performing some calculation.) Everything here can be written as a code line, that is, as a line of instructions. Any Turing machine is described by a finite set of such lines, say, L_1, L_2, L_3, L_4, L_5. Yes, it looks like a computer program, and is in fact a computer program. Each line L_i can be written in binary code, and we can translate the instructions in any computer program into that kind of Turing machine program and vice versa.

One important point: Turing machines can enter infinite loops. Suffices to order it to make a never ending calculation, like e.g. the division of, say, 1 by 0, or perhaps constructing a machine that moves to the right and then to the left one square, forever. The old electromechanical calculating machines would only stop when doing that division if we turned off its power; modern hand-held electronic calculators cut off such known never ending procedures. This is an example of an infinite loop a Turing machine can enter into, but as we'll soon see not all such infinite loops can be predicted — at least by some general computer program.

The universal machine

Therefore we can code every Turing machine program as an integer (which will be very large for programs like the ones we use in our computers). And there will be a computable way of listing them; the listing of all Turing machines is done by another Turing machine, the universal machine.

Our computers are versions of that universal machine. The universal machine allows itself to be programmed, that is, given the numerical code for a given program, it simulates the operation of that program; any program, in fact. (The chief difference between Turing machines and our concrete, everyday computers is that Turing machines are supposed to have a potentially infinite memory, while the size of our computers' RAM and hard disk memories is of course limited.)

The halting problem

We'll do some math in this section, but before we get our hands dirty let's tell a nearly folklore-like tale about the halting problem. In the 1950s computers were mainly built by electrical engineers, and once upon a time a team of engineers who were working on one of the brand-new "electronic brains" of the period met some colleagues from the math department at some university's cafeteria. The engineers began to discuss their work with their brand new computer and said that they were trying to develop a kind of test program that would avoid bugs: the test program would know about the program they were running in the machine and out of that knowledge it would test beforehand any input to see whether it resulted in an infinite loop — a never ending succession of operations without any output — or not.

So far they hadn't been successful.

Then one of the math department people chuckled and started to laugh. What are you laughing about? complained the engineers. The math guy's answer: what you are trying to do is impossible; Turing proved it in 1936, two decades ago.

You are trying to write a program that solves the halting problem, and that cannot be done.

Why can't it be done?

We present a short argument that shows why such a program cannot exist:

- List all Turing machines $M_0, M_1, M_2, \ldots$.

- Now suppose that there is a program $g(x, y)$ that performs as follows:

  - $g(x, y) = 1$ if and only if machine of program y stops over input x and gives some output.
  - $g(x, y) = 0$ if and only if machine of program y enters an infinite loop when it receives input x.

- If g is a program that can be constructed then h which we now describe can also be constructed:

- $h(x) = 1$ if and only if $g(x, x) = 0$.
- $h(x)$ diverges if and only if $g(x, x) = 1$.
  (In order to make h enter an infinite loop we can plug a diverging subroutine to it at convenient places.)

- Now let's go back to our listing of all Turing machines $M_0, M_1, M_2, \ldots$. If h is a program, it can be implemented as a Turing machine, and there is a k so that $M_k = h$.

- Then:

  - If $h(k) = M_k(k) = 1$ we get that $g(k, k) = 0$, from the definition of h. Now, from the definition of g, $M_k(k)$ must diverge. A contradiction.
  - If $h(k) = M_k(k)$ diverges, then $g(k, k) = 1$, which means that $M_k(k)$ converges. Another contradiction.

- Therefore we can neither write a program like g nor one like h.

Gödel's first incompleteness theorem revisited

The argument we now present is quite brief. Suppose that we can formalize in our theory S (with enough arithmetic) the sentence "Turing machine M_k over input m diverges." (Yes, we can do it.)

Suppose that there is a proof of all such sentences in S, any k, m.

If there is one such proof procedure, then we can make it into a calculation procedure. Thus we would settle all non-halting instances of the halting problem for k, m.

That's impossible. Therefore there are k_0, m_0 so that while it is true that machine k_0 over m_0 diverges, we cannot prove within S that in fact machine k_0 over input m_0 diverges. Also, as we've supposed that S doesn't prove false assertions, it cannot prove the negation of the sentence:

"*Turing machine M_{k_0} over input m_0 diverges.*"

Thus that sentence is undecidable within S. And it is an undecidable sentence with an obvious mathematical meaning. This proof of Gödel's first incompleteness theorem is due to Emil Post. It shows the interdependence between undecidability (the unsolvability of the halting problem, in this case) and incompleteness (the existence of sentences which one can neither prove nor disprove within formal systems like S).

The Church–Turing thesis

We sketched above a few criteria that try to describe what we do when we perform a computation. A computation is essentially a finitistic, deterministic procedure that allows us to obtain some output from a given finite set of inputs. It is a kind of cooking recipe for a calculation.

But the criteria we have presented are informal. They become formal when we write them down in some kind of mathematical language, as we have done with Turing machines. How can we make sure that our conceptions about how to perform a computation are mirrored in some formalized kind of prescription for it?

We can't. Thus the so-called *Church–Turing thesis*. Roughly, it can be formulated as:

> *If we can write down an algorithm (a computation procedure) for some calculation, then it can be formulated as a Turing machine.*

Turing wrote his PhD thesis under the guidance of logician Alonzo Church (1903–1995) and passed it at Princeton in 1938. Church had just proposed that the informal concept of computation should be formalized in his λ-calculus. It was shown in 1936 that Church's λ-calculus and Kleene's general recursive functions described the same set of functions, and soon after the presentation of Turing's formalization, it was proved that again they described the same set of functions. Another equivalent procedure was proposed by Emil Post in 1939; more recent formalizations are the Markov algorithms and cellular automata (proved to be equivalent to Turing machines in 1971).

So most logicians tend to accept a kind of orthodox viewpoint and equate the informal notion of computation to those formalized schemes (Turing machines, general recursive functions, λ-calculus, and the like).

But should we thus restrict our concept of computation? If you wish to listen to the case of hypercomputation, go to the last chapter, where we discuss it at length.

Diophantine equations; Hilbert's 10th problem

Consider a simple equation like:

$$x^2 + y^2 = z^2. \tag{3}$$

This equation has integer solutions such as $x = 3$, $y = 4$ and $z = 5$. Now consider the similar equation:

$$x^3 + y^3 = z^3. \tag{4}$$

As it is well-known, now we have no nontrivial integer solutions, that is, solutions beyond 0 and 1. That new equation is an instance of Fermat's Last Theorem, and both are examples of Diophantine equations.

Write down a polynomial on a finite number of variables and choose integer coefficients for the monomials that add up it it. Put it equal to zero,

$$p(x, y, z, t, v, \ldots, w) = 0.$$

If we ask for integer solutions we have another, now a general example of a Diophantine equation. Hilbert's 10th problem is:

Given a Diophantine equation:

$$p(x,y,z,t,v,\ldots,w) = 0, \tag{5}$$

is there any procedure that in a finite number of steps allows us to determine whether it has solutions in the integers?

The answer: no.

Four authors contributed to the solution of Hilbert's Tenth Problem, Martin Davis, Julia Robinson, Hilary Putnam and Yuri Matiyasevich. Emil Post suggests in 1944 that "Hilbert's Tenth Problem begs for an unsolvability solution." Martin Davis gives a first contribution towards the solution in 1949, followed by a conjecture by Julia Robinson in 1950 that essentially would lead to the solution of the question. Putnam enters the fray in 1959 together with Davis and with the help of an unproved conjecture on prime numbers they zero in the solution. The final step is given by Yuri Matiyasevich in 1970.

The solution of Hilbert's Tenth Problem can be formulated for our purposes as: let us be given a Turing machine of code (program) described by the natural number n, that is, M_n. Suppose that its input is given by k, also a natural number. We can construct a Diophantine polynomial $p(\langle n,k\rangle x,y,z,\ldots)$ so that:

Turing machine M_n stops over input k and gives some output if and only if $p(\langle n,k\rangle x,y,\ldots) = 0$ has integer solutions.

Moreover there is a step-by-step, computational procedure that allows us to obtain p out of machine M_n.

We get several extra bonuses out of the solution of Hilbert's Tenth Problem:

- *Existence of an universal polynomial.* There are universal Turing machines, that is, Turing machines that simulate any particular Turing machine. Similarly there are universal Diophantine polynomials, that is, polynomials that represent the universal machines and therefore that are such that they may stand for any particular Turing machine.

  The universal polynomial may be quite a monster: if it has 9 variables, it degree will reach 1.6×10^{45}. However if we build it with 58 variables, then its degree falls drastically to 4.

- *The prime number function.* An old conundrum in prime number theory was the search for a function that enumerates all prime numbers. That function results from the following result:

  There is a polynomial with integer coefficients which ranges over the nonnegative integers whose positive values are the prime numbers.

- *More results about sets of numbers.* There are polynomials that enumerate the Fibonacci numbers (this one has 2 variables and is of degree 5), Mersenne primes (one of them has degree 26 and 13 variables) and even perfect numbers (one of them has 13 indeterminates and degree 27).

- *Complexity of mathematical proofs in terms of sums and products.* That's another consequence of the existence of an universal polynomial:

  > *Given theory S and some formal sentence p in it, if S proves p then there is a "computational proof" of p with 100 additions and products of integers.*

  That is, we translate the inner workings of S into the universal polynomial, and the proof of p becomes the check that the corresponding polynomial is zero, that is the machine M_S that simulates the proofs of S stops over input p adequately coded.

The solution of Hilbert's Tenth Problem shows that to talk about computation in general is to talk about Diophantine equations, and vice-versa. And it provides an easy and natural way to axiomatize computer science. It is enough to pick up some theory like our S that contains arithmetic an get the mirror image of Turing machines and the like within Diophantine equation theory.

Undecidable issues

However there is a catch here: Kleene's function F. That function is obviously computable and total — intuitively computable and total, we may say — but our theory S cannot prove that a corresponding F_S (the subscript is to stress the relation to theory S) it is total. Now if we consider theory $S' = S + (F_S$ is total) then there will be a function $F_{S'}$ that again cannot be proved total in the extended theory, and so on, forever.[14]

This means: suppose that we manage to formulate a totally different axiomatic theory for computer science which however is based on classical logic (the classical predicate calculus), has enough arithmetic and has a computable list of theorems. Then it will exhibit the same incompleteness phenomenon as S and there will be a mystery function like F in that new theory.

Function F_S and the Busy Beaver function

Tibor Radó (1895–1965) was a hungarian-american mathematician that offered a powerful counterexample to the belief that mathematicians do nothing of interest when over 40. He published his most famous and celebrated paper, "On non-computable functions," when he was 67.

(Radó's was a very adventurous life. He fought in World War I for Hungary, was captured by the Russian tzarist troops and sent to a prisoners' camp in Siberia, from which he escaped and managed to go back to Hungary. He moved to the United States in 1929.)

There are several ways we can define the Busy Beaver function, or Busy-Beaver-like functions, for there are several such functions which share the main

---

[14]There will be an infinite bound in that sequence, however. More about it later in this chapter.

properties stressed by Radó, namely, being clearly definable, noncomputable, and dominating all total computable functions. One of them goes as follows:

> *Consider all n-state Turing machines which only print zeros and ones. Then the Busy Beaver function $\Sigma(n)$ is the largest number of 1s $+1$ which a n-state machine that starts over a blank tape prints on its tape and stops.*

As Radó stresses in his paper, construction of the Busy Beaver function is based on a simple principle: every finite set of positive integers has a maximum element. There is a finite number of *n*-states Turing machines, and since there can only be a finite number of *n*-state machines, there necessarily will be one that prints the maximum number of 1s on the tape and stops.

The Busy Beaver function is noncomputable. Radó proves it by showing that it tops all total computable functions. Radó's argument is algebraically developed but we can easily see that it has the required properties. We will use a slightly modified Busy Beaver function:

> *The modified Busy Beaver function $\Sigma'(n)$ is the number of 1s a n-state Turing machine prints when placed over a blank square on the machines tape, or over 0 or 1 or 2 or . . . n.*

Clearly $\Sigma'(n) \geq \Sigma(n)$.

- The Busy Beaver function tops all total computable functions. This is easy to see: the total functions will appear among the ones we must take into account for the Busy Beaver computation as long as we increase the value of *n*.

- Then as it tops all total computable functions, it cannot be computable itself.

Now go back to p. 6 and check the way we defined F (or F_S). We start out of a listing of total computable functions which are proved to have that property in S and obtain for F a different value of the function at step *n* by the definition $F_S(n) = f_n(n) + 1$. However we could proceed as follows:

- Start the listing of the theorems.
- Pick up those that say: "function f is total computable."
- Out of that we can build a list $f_0, f_1, f_2, \ldots$, of total computable functions with their values like that:

$$f_0(0), f_0(1), f_0(2), f_0(3), \ldots$$
$$f_1(0), f_1(1), f_1(2), f_1(3), \ldots$$
$$f_2(0), f_2(1), f_2(2), f_2(3), \ldots$$
$$f_3(0), f_3(1), f_3(2), f_3(3), \ldots$$
$$\ldots$$

- Now put: $F_S(n)$ is the largest value among those attained by functions $f_0, f_1, \ldots$, up to *n*, plus one.

The definition of this new F_S is similar to the Busy Beaver. With one single difference: F_S is a computable function, as we can compute all values for the $_0$, $f_1, \ldots$, up to n and beyond. So, F_S is a kind of Busy Beaver function[15] as seen from within a formal theory S.

So, we have a very important property:

> *No theory like S can prove that (what it sees) the Busy Beaver function is total (that is, defined for all values $0, 1, 2, \ldots, n \ldots$).*

That's immediate, for the Busy Beaver function grows beyond all computable functions that can be proved total by S. Now what happens if we build a sequence of theories such as:

$$S_0 = S.$$

$$S_1 = S_0 + (F_{S_0} \text{ is total.})$$

$$S_2 = S_1 + (F_{S_1} \text{ is total.})$$

$$\ldots$$

We'll soon see the meaning of that progression of theories.

A final note on the Busy Beaver, which will only be fully clarified in the next section. Given each value n there are algorithms A_n and A'_n that compute the Busy Beaver functions $\Sigma(n)$ and Σ'_n. However the sequence $A_0, A_1, A_2, \ldots$ and the corresponding one $A'_0, A'_1, \ldots$ do not add up to an algorithm for the whole of the Busy Beaver function, for those sequences cannot be generated by computer programs. They are irreducibly chaotic in Chaitin's sense.

Busy-Beaver-like functions

The Busy Beaver function isn't as artificial as it looks. Such noncomputable, fast-growing functions appear in several areas of mathematics:

- *Variants of the Busy Beaver function.* If you go through the literature, you'll notice that there actually are two fast-growing functions related to what Radó called "the Busy Beaver game," and those functions are our $\Sigma(n)$ but also a related $S(n)$, which we are not going to discuss here.

  There are also many modified formulations of the Busy Beaver function, which give different mathematical objects which keep the main characteristics of the original BB function Σ: its fast-growing nature and its uncomputability.

- *Chaitin's $e(n)$.* Chaitin introduced in 1972 a function $e(n)$ that bounds all proofs in an axiomatic theory like our much-abused S. If we could compute $e(n)$ we would be able to decide theoremhood in S — we can list all the theorems of a theory like S, but we have no decision procedure to check for an arbitrary formal sentence whether it is a theorem of S or not. $e(n)$ is related to Chaitin's wonder constant Ω (more about Ω later).

---

[15]Doria owes that remark to Leo Gordeev.

- *The counterexample function to P vs. NP.* That one will also be discussed in the next chapters. It is a highly oscillating noncomputable function that lists all counterexamples to $P = NP$, that is all failures of fast algorithms to solve all instances of a problem in class NP. In its peaks it overtakes all intuitively total computable functions.

Turing 1939: progressions of theories

Alan Turing passed his PhD thesis at Princeton in mid-1938 under the guidance of Alonzo Church and then had it published by the Proceedings of the London Mathematical Society, the same journal that had published the paper on mathematical machines.

But the issue here is a different one. Turing starts from Gödel incompleteness and wishes to see what can be done to obtain a complete axiomatic system, that is, one where truth and provability are a perfect match.

How would that be possible?

We will now sketch a delicate argument, first explored by Turing but only fully developed a quarter of a century later by Solomon Feferman. It deals with "progressions of theories," already mentioned in the previous section.

Let's start at a slow pace. We've mentioned that Gödel's undecidable sentence G is what is called a Π_1 sentence (see the discussion in p. 8). It's a sentence of the form "for every x, x has the property P," where we can decide whether in fact x has P by computational means.

We have also mentioned that $\mathrm{Con}(S)$, the formalized sentence that asserts the consistency of S, is equivalent to G. Thus theory $S_1 = S_0 + \mathrm{Con}(S_0)$ proves lots of Π_1 sentences, beginning with Gödel's undecidable sentence G_0 (the 0 index stresses that it refers to theory S_0). Of course once S_0 is supposed consistent, then so is S_1.

Now let's take a look at $S_2 = S_1 + \mathrm{Con}(S_1)$. S_2 is consistent, and proves the Gödel undecidable sentence for S_1, that is G_1 as well as infinitely many new Π_1 sentences. And so on? Do we exhaust all true Π_1 sentences in that process? Yes — it suffices to iterate it to the so-called constructible ordinal $\omega + 1$.

The halting problem revisited

Each true non-halting instance of a Turing machine is represented by a Π_1 sentence. Actually there is the following result:

> If $h_1, h_2, \ldots, h_k$ are true nonhalting instances of the halting problem for a given universal Turing machine, then there is an algorithm $A(h_1, \ldots, h_k)$ that settles those nonhalting instances.

That is, the algorithm decides that the prescribed universal machine will not stop at the given instances. Proof is very simple: theory $S + h_1 + \cdots + h_k$ proves that the given machine doesn't halt for those instances. (Of course one has to use ingenuity in order to obtain that theory and the algorithm that follows from it.)

This explains why there always is an algorithm to compute the Busy Beaver function for any state n, while we cannot collect the particular algorithms into a single, all-encompassing superalgorithm that summarizes all procedures to compute all values of the Busy Beaver. Similarly, there is no all-encompassing algorithm to settle all instances of the halting problem, while we can obtain them for specific instances.

Beyond the Gödel phenomenon

Do we have axioms for arithmetic that give us a complete theory? Yes. We'll see that in the next chapters. (Recall that for a complete theory truth is proof, proof is truth: everything that is provable is true, and vice-versa.)

Do we get anything useful out of those theories? Perhaps. Wait and see.

2. Complexity, Randomness

I N 1931 GÖDEL DISCOVERED INCOMPLETENESS by considering "This statement is unprovable!" As we said in the previous chapter, the work of Turing in 1936 revealed that incompleteness is a corollary of a deeper phenomenon, uncomputability. In this chapter we're going to go beyond Turing by considering an extreme form of uncomputability, namely algorithmic irreducibility and randomness.[16]

We'll study the halting probability Ω, an extreme case of maximal uncomputability, a place in pure math where God seems to play dice. Imagine writing the numerical value of Ω in base-two:

$$\Omega = .1101011\ldots \tag{1}$$

The 0, 1 bits of the numerical value of Ω are algorithmically and logically irreducible, they appear to be completely haphazard, without any structure. Each bit of Ω has got to be a 0 or a 1, but these are mathematical facts that are true for no reason, more precisely, for no reason simpler than themselves. They seem more like contingent truths than like necessary truths. And, as we shall see, Ω's irreducible complexity can be found throughout pure mathematics, even in such traditional fields as number theory and algebra.

In fact, as is discussed in this chapter, the reason that the bits of Ω seem to be completely accidental and random is because they are an irredundant compression of the answers to all possible instances of the halting problem. Once all the redundancy has been compressed out of something, what is left seems completely random, but is in fact packed full of useful information. Such is the case with the bits of Ω. Ω is the best way to pack all the answers to individual cases of the halting problem into a single real number.

The halting probability Ω plays a leading role in this chapter, but our goals are actually much broader. In this chapter we present an information-theoretic perspective on epistemology using software models. We shall use the notion of algorithmic information to discuss what is a physical law, to determine the limits of the axiomatic method, and to analyze Darwin's theory of evolution.

And the best way to understand the deep concepts of conceptual complexity and algorithmic information, which are the basis for everything in this chapter, is to see how they evolved, to know their long history. Let's start with Hermann Weyl and the great philosopher/mathematician G. W. Leibniz. That everything that is true is true for a reason is rationalist Leibniz's famous *principle of sufficient reason*. The bits of Ω seem to refute this fundamental principle and also the idea that everything can be proved starting from self-evident facts.

---

[16]Portions of this chapter first appeared in G. J. Chaitin, "Metaphysics, metamathematics and metabiology," in H. Zenil, *Randomness Through Computation*, World Scientific (2011).

Weyl, Leibniz, complexity and the principle of sufficient reason

What is a scientific theory?

The starting point of algorithmic information theory, which is the subject of this chapter, is this toy model of the scientific method:

theory/program/010 → **Computer** → experimental data/output/110100101.

A scientific theory is a computer program for exactly producing the experimental data, and both theory and data are a finite sequence of bits, a bit string. Then we can define the complexity of a theory to be its size in bits, and we can compare the size in bits of a theory with the size in bits of the experimental data that it accounts for.

That the simplest theory is best, means that we should pick the smallest program that explains a given set of data. Furthermore, if the theory is the same size as the data, then it is useless, because there is always a theory that is the same size as the data that it explains. In other words, a theory must be a compression of the data, and the greater the compression, the better the theory. Explanations are compressions, comprehension is compression!

Furthermore, if a bit string has absolutely no structure, if it is completely random, then there will be no theory for it that is smaller than it is. Most bit strings of a given size are incompressible and therefore incomprehensible, simply because there are not enough smaller theories to go around.

This software model of science is not new. It can be traced back via Hermann Weyl (1932) to G. W. Leibniz (1686)! Let's start with Weyl. In his little book on philosophy *The Open World: Three Lectures on the Metaphysical Implications of Science*, Weyl points out that if arbitrarily complex laws are allowed, then the concept of law becomes vacuous, because there is always a law! In his view, this implies that the concept of a physical law and of complexity are inseparable; for there can be no concept of law without a corresponding complexity concept. Unfortunately he also points out that in spite of its importance, the concept of complexity is a slippery one and hard to define mathematically in a convincing and rigorous fashion.

Furthermore, Weyl attributes these ideas to Leibniz, to the 1686 *Discours de métaphysique*. What does Leibniz have to say about complexity in his *Discours*? The material on complexity is in Sections V and VI of the *Discours*.

In Section V, Leibniz explains why science is possible, why the world is comprehensible, lawful. It is, he says, because God has created the best possible, the most perfect world, in that the greatest possible diversity of phenomena are governed by the smallest possible set of ideas. God simultaneously maximizes the richness and diversity of the world and minimizes the complexity of the ideas, of the mathematical laws, that determine this world. That is why science is possible!

A modern restatement of this idea is that science is possible because the world seems very complex but is actually governed by a small set of laws having low conceptual complexity.

And in Section VI of the *Discours*, Leibniz touches on randomness. He points out that any finite set of points on a piece of graph paper always seems to follow a law, because there is always a mathematical equation passing through those very points. But there is a law only if the equation is simple, not if it is very complicated. This is the idea that impressed Weyl, and it becomes the definition of randomness in algorithmic information theory.[17]

Finding elegant programs

So the best theory for something is the smallest program that calculates it. How can we be sure that we have the best theory? Let's forget about theories and just call a program *elegant* if it is the smallest program that produces the output that it does. More precisely, a program is elegant if no smaller program written in the same language produces the same output.

So can we be sure that a program is elegant, that it is the best theory for its output? Amazingly enough, we can't: It turns out that any formal axiomatic theory A can prove that at most finitely many programs are elegant, in spite of the fact that there are infinitely many elegant programs. More precisely, it takes an N-bit theory A, one having N bits of axioms, having complexity N, to be able to prove that an individual N-bit program is elegant. And we don't need to know much about the formal axiomatic theory A in order to be able to prove that it has this limitation.

What is a formal axiomatic theory?

All we need to know about the axiomatic theory A, is the crucial requirement emphasized by David Hilbert that there should be a proof-checking algorithm, a mechanical procedure for deciding if a proof is correct or not. It follows that we can systematically run through all possible proofs, all possible strings of characters in the alphabet of the theory A, in size order, checking which ones are valid proofs, and thus discover all the theorems, all the provable assertions in the theory A.[18]

That's all we need to know about a formal axiomatic theory A, that there is an algorithm for generating all the theorems of the theory. This is the software model of the axiomatic method studied in algorithmic information theory. If the software for producing all the theorems is N bits in size, then the

---

[17]*Historical Note:* Algorithmic information theory was first proposed in the 1960s by R. Solomonoff, A. N. Kolmogorov, and G. J. Chaitin. Solomonoff and Chaitin considered this toy model of the scientific method, and Kolmogorov and Chaitin proposed defining randomness as algorithmic incompressibility.

[18]*Historical Note:* The idea of running through all possible proofs, of creativity by mechanically trying all possible combinations, can be traced back through Leibniz to Ramon Llull in the 1200s.

complexity of our theory A is defined to be N bits, and we can limit A's power in terms of its complexity $H(A) = N$. Here's how:

Why can't you prove that a program is elegant?

Suppose that we have an N-bit theory A, that is, that $H(A) = N$, and that it is always possible to prove that individual elegant programs are in fact elegant, and that it is never possible to prove that inelegant programs are elegant. Consider the following paradoxical program P:

> *P runs through all possible proofs in the formal axiomatic theory A, search-*
> *ing for the first proof in A that an individual program Q is elegant for*
> *which it is also the case that the size of Q in bits is larger than the size*
> *of P in bits. And what does P do when it finds Q? It runs Q and then P*
> *produces as its output the output of Q.*

In other words, the output of P is the same as the output of the first provably elegant program Q that is larger than P. But this contradicts the definition of elegance! P is too small to be able to calculate the output of an elegant program Q that is larger than P. We seem to have arrived at a contradiction!

But do not worry; there is no contradiction. What we have actually proved is that P can never find Q. In other words, there is no proof in the formal axiomatic theory A that an individual program Q is elegant, not if Q is larger than P. And how large is P? Well, just a fixed number of bits c larger than N, the complexity $H(A)$ of the formal axiomatic theory A. P consists of a small, fixed main program c bits in size, followed by a large subroutine $H(A)$ bits in size for generating all the theorems of A.

The only thing tricky about this proof is that it requires P to be able to know its own size in bits. And how well we are able to do this depends on the details of the particular programming language that we are using for the proof. So to get a neat result and to be able to carry out this simple, elegant proof, we have to be sure to use an appropriate programming language. This is one of the key issues in algorithmic information theory, which programming language to use, and I'll talk about that later.[19]

Farewell to reason: The halting probability Ω[20]

So there are infinitely many elegant programs, but there are only finitely many provably elegant programs in any formal axiomatic theory A. The proof of this is rather straightforward and short. Nevertheless, this is a fundamental information-theoretic incompleteness theorem that is rather different in style

---

[19]See also the chapter on "The Search for the Perfect Language" in Chaitin, *Mathematics, Complexity and Philosophy*, in press.

[20]*Farewell to Reason* is the title of a book by Paul Feyerabend, a wonderfully provocative philosopher. We borrow his title here for dramatic effect, but he does not discuss Ω in this book or any of his other works.

from the classical incompleteness results of Gödel, Turing and others that were discussed in the previous chapter.

An even more important incompleteness result in algorithmic information theory has to do with the halting probability Ω, the numerical value of the probability that a program p whose successive bits are generated by independent tosses of a fair coin will eventually halt:

$$\Omega = \sum_{p \text{ halts}} 2^{-(\text{size in bits of } p)}. \tag{2}$$

To be able to define this probability Ω, it is also very important how you chose your programming language. If you are not careful, this sum will diverge instead of being ≤ 1 like a well-behaved probability should.

Turing's fundamental result is that the halting problem in unsolvable. In algorithmic information theory the fundamental result is that the halting probability Ω is algorithmically irreducible or random. It follows that the bits of Ω cannot be compressed into a theory less complicated than they are. They are irreducibly complex. It takes N bits of axioms to be able to determine N bits of the numerical value

$$\Omega = .1101011\ldots \tag{3}$$

of the halting probability. If your formal axiomatic theory A has $H(A) = N$, then you can determine the values and positions of at most $N + c$ bits of Ω.

In other words, the bits of Ω are logically irreducible, they cannot be proved from anything simpler than they are. Essentially the only way to determine what are the bits of Ω is to add these bits to your theory A as new axioms. But you can prove anything by adding it as a new axiom. That's not using reasoning!

So the bits of Ω refute Leibniz's principle of sufficient reason: they are true for no reason. More precisely, they are not true for any reason simpler than themselves. This is a place where mathematical truth has absolutely no structure, no pattern, for which there is no theory!

Adding new axioms: Quasi-empirical mathematics[21]

So incompleteness follows immediately from fundamental information - theoretic limitations. What to do about incompleteness? Well, just add new axioms, increase the complexity $H(A)$ of your theory A! That is the only way to get around incompleteness.

In other words, do mathematics more like physics, add new axioms not because they are self-evident, but for pragmatic reasons, because they help mathematicians to organize their mathematical experience just like physical theories help physicists to organize their physical experience. After all,

---

[21]The term *quasi-empirical* is due to the philosopher Imre Lakatos, a friend of Feyerabend. For more on this school, including the original article by Lakatos, see the collection of quasi-empirical philosophy of math papers edited by Thomas Tymoczko, *New Directions in the Philosophy of Mathematics*.

Maxwell's equations and the Schrödinger equation are not at all self-evident, but they work! And this is just what mathematicians have done in theoretical computer science with the hypothesis that $P \neq NP$, in mathematical cryptography with the hypothesis that factoring is hard, and in abstract axiomatic set theory with the new axiom of projective determinacy.[22]

This concludes the introductory historical overview. Now let's start over and do things a little more carefully.

Defining information content and conceptual complexity

Here is how we define *information content* or *conceptual complexity* for finite and infinite objects.

For the sake of brevity, I will call a Turing machine a *computer*, and will use $C(p)$ to denote the output produced by running the program p on the computer C. We are only interested in *binary* computers, so the p that we consider will always be bit strings, and we shall try to minimize the size in bits $|p|$ of p.

Pick an *optimal self-delimiting universal binary computer* U to use as our standard, official programming language. U has the property that for each self-delimiting binary computer C there exists a prefix π_C such that

$$U(\pi_C p) = C(p), \tag{4}$$

that is to say, U carries out the same computation that C does if U is given the program p for C preceded by a fixed prefix π_C that depends on C but not on p. Thus, programs for U are at most a fixed number of bits larger than programs for C, at most $|\pi_C|$ bits longer, in fact, which is the size in bits of a self-delimiting interpreter π_C for U to simulate C.

Self-delimiting means that π_C indicates within itself how long it is, so that when it is concatenated with the program p it is nevertheless possible for U to decide where π_C ends and p begins. The entire programs $\pi_C p$ and p for U and C respectively, are similarly required to be self-delimiting, in that U and C can realize where $\pi_C p$ and p end without have to read beyond the ends of their respective binary programs.

Henceforth we use a particular U to measure the size of programs, to define the program-size or conceptual complexity H. The choice of U does not affect things too much, since two optimal self-delimiting universal binary computers U and U' require programs of almost the same size.

For individual digital objects, $H(x) \equiv$ the size in bits of the smallest self-delimiting program p that calculates x and then halts. I.e.,

$$H(x) \equiv \min_{U(p)=x} |p|. \tag{5}$$

For infinite sets X of digital objects, $H(X) \equiv$ the size in bits of the smallest self-delimiting program p that runs forever outputting all the elements of X

---

[22]See the article on "The Brave New World of Bodacious Assumptions in Cryptography" in the March 2010 issue of the *AMS Notices*, and the article by W. Hugh Woodin on "The Continuum Hypothesis" in the June/July 2001 issue of the *AMS Notices*.

(and only the elements of X). So $H(X)$ is the size in bits of the smallest self-delimiting program p such that $U(p) = X$.

As we said before, most bit strings and binary sequences have maximal complexity and are called *algorithmically irreducible* or *algorithmically random*. Most N-bit strings x have

$$H(x) \approx N + H(N). \tag{6}$$

And if the bits of an infinite binary sequence X are chosen using independent tosses of a fair coin, then with probability one there exists a constant c such that

$$H(\text{the first } N \text{ bits of } X) > N - c$$

for all N. This implies Martin-Löf statistical randomness, namely that X is not contained in any *constructive measure zero set*.[23] Amazingly enough, it turns out that Martin-Löf randomness is actually equivalent to algorithmic irreducibility, the fact that

$$H(\text{the first } N \text{ bits of } X) > N - c, \tag{7}$$

even though these concepts are defined completely differently. There is also a lesser-known but technically superior measure-theoretic definition of statistical randomness due to Robert Solovay, which also turns out to be equivalent to algorithmic irreducibility. For a proof of these equivalence theorems, see Chaitin, *Exploring Randomness*.

Furthermore, the fact that programs are self-delimiting enables us to define the halting probability Ω as a sum over all programs that halt when run on U:

$$0 < \Omega \equiv \sum_{p \text{ halts}} 2^{-|p|} < 1. \tag{8}$$

Each program p that halts contributes $1/2^{(p\text{'s size in bits})}$ to the halting probability Ω, since each bit of p is chosen using an independent toss of a fair coin. Note that the precise numerical value of $\Omega = \Omega_U$ depends on the choice of universal machine U, but Ω's remarkable properties do not. If U were not self-delimiting, this sum would diverge. If U were not optimal, Ω would not be maximally complex and statistically random.

Since Ω is algorithmically irreducible, which we shall show later in this chapter, it is also Martin-Löf random and therefore Borel normal. In other words, written in any base b, the b different base-b digits of Ω all provably have equal limiting relative frequency $1/b$, and each block of k consecutive base-b digits has limiting relative frequency $1/b^k$.[24] Ω is the only natural example of a real number that is provably Borel normal for every base b and for blocks of every size k, even though Borel proved that for a real number

---

[23] A set of real numbers is a constructive measure zero set if it can be covered by an arbitrarily small covering and this can be done algorithmically. The infinite sequences X correspond to the reals $x \in [0,1] \equiv \{x : 0 \leq x \leq 1\}$; simply place a decimal point in front of the bits of X to get the base-two representation for x.

[24] For a self-contained proof that Ω is Borel normal, see Chaitin, *Algorithmic Information Theory*.

$x \in [0,1] \equiv \{x : 0 \le x \le 1\}$ this is true with probability one. It is conjectured that the real number $\pi = 3.1415926\ldots$ also has this property, but π is not algorithmically irreducible. In fact,

(The first N digits of π) is a computable function f_π of N.

And so

$$H(\text{the first } N \text{ digits of } \pi) = H(f_\pi(N)) \le H(f_\pi) + H(N) = c + H(N) \approx \log_2 N.$$

Why theories? Subadditivity and mutual information

It is crucial that because we are using self-delimiting programs, the information content of two objects (of computing the pair) is bounded by the sum of the individual information contents.

For x and y individual digital objects:

$$H(x,y) \le H(x) + H(y) + c. \tag{9}$$

Here we can simply concatenate the two self-delimiting programs, and then run one and then the other. The constant c is the size in bits of the prefix π_C that makes U do this. In other words, there is a C such that

$$C(x^*y^*) = \text{the pair } (x,y). \tag{10}$$

Here x^* is a minimum-size program for x and y^* is a minimum-size program for y. Therefore

$$U(\pi_C x^* y^*) = \text{the pair } (x,y). \tag{11}$$

And

$$H(x,y) \le |\pi_C x^* y^*| = |\pi_C| + H(x) + H(y) = H(x) + H(y) + c. \tag{12}$$

Similarly, for infinite sets of digital objects X and Y:

$$H(X \cup Y) \le H(X) + H(Y) + c'. \tag{13}$$

Here we have to imagine both of these infinite computations proceeding in parallel, and interleave the bits of their respective programs in the order that they are required.

In particular, consider formal mathematical theories, that is, infinite sets of theorems, X and Y. Then $H(X)$ is the number of bits of axioms in the theory X, in other words, the size in bits of the proof-checking algorithm, more precisely, the size in bits of the smallest program for running through all possible proofs checking which ones are correct and finding all the theorems. $H(X)$ is the best possible way of doing this.

Consider a formal axiomatic theory X. Our crucial theorem on proving that programs are elegant can now be stated more precisely like this:

> *If X has the property that "p is elegant" $\in X$ only if p is elegant, then "p is elegant" $\in X$ only if $|p| \leq H(X) + c$.*

In other words, to prove that an N-bit program is elegant you need at least N bits of axioms. Similarly:

> *If X has the property that "The kth bit of Ω is a 0" and "The kth bit of Ω is a 1" are in X only if they are true, then X can determine at most $H(X) + c'$ bits of Ω.*

In other words, an N-bit theory can determine at most N bits of Ω.[25]

How much information do x and y or X and Y have in common?

Here is how we measure the *mutual information* for individual objects:

$$H(x : y) \equiv H(x) + H(y) - H(x, y), \tag{14}$$

and for infinite sets of objects:

$$H(X : Y) \equiv H(X) + H(Y) - H(X, Y). \tag{15}$$

Low mutual information is *algorithmic independence*:

$$H(x, y) \approx H(x) + H(y), \quad H(X, Y) \approx H(X) + H(Y). \tag{16}$$

The mutual information is the extent to which it is simpler to compute them together than to compute them separately, and in the case of independent objects there is no essential difference between computing them together and computing them separately.

For example, if you choose two N-bit strings x, y at random, with high probability all they have in common is their length:

$$H(x : y) \approx H(N) \approx \log_2 N. \tag{17}$$

This information-theoretic notion of independence is connected both with statistical and with logical independence.

Let me show you the connection with logical independence.

The reason that theories work is that things are not all independent; there is a lot of mutual information, there are a lot of common principles. In particular, let's consider mathematical theories:

> *The reason that theories work is that there are common principles, that there is a lot of mutual information between the theorems. Otherwise we would just have to add each theorem as an independent axiom!*

---

[25] We shall not prove this here, but you can see this by combining ideas from our proof that only finitely many programs are provably elegant with the fact that Ω is algorithmically irreducible, which we shall prove later in this chapter. For a complete proof that is worked out in detail using LISP, see Chaitin, *The Limits of Mathematics*.

So our information-theoretic, complexity viewpoint suggests that physics and mathematics are not that different. In both cases, theories work for roughly the same reason: they are compressions, they find common principles. And this suggests a quasi-empirical view of mathematics, and adding non-self-evident new principles for pragmatic reasons, because they help us to organize our mathematical experience, just like scientific theories help us to organize our physical experience.

Combining theories and making conjectures

Following the late Ray Solomonoff, one of the founders of algorithmic information theory, let's show how to combine alternative physical or mathematical theories in order to make predictions.

Say we have several, for example, two theories, which explain what we have already seen and predict different future events. E.g., say we have seen 0110 and we want to know the next bit, and we have two theories, two computer programs that output 0110 and that then continue with other bits. At the start of this chapter, we said the simplest theory is best, but what if we have several theories and want to take them all into account, what then?

So we have two p with $U(p) = 0110x$ and how much weight should we put on each prediction x? According to Solomonoff, each p gets weight $2^{-|p|}$, 1/(two raised to the size in bits of p). So if one p is three bits long, and the other is five bits long, and both explain what we have already seen and predict different continuations, then these weights will be 1/8 and 1/32. So if the first predicts a 0 and the second theory predicts a 1, then to get probabilities from these weights we must normalize them:

$$\frac{1/8}{1/8 + 1/32}, \frac{1/32}{1/8 + 1/32}$$

are the respective weights for the 0 and 1 predictions.[26]

Similarly in the case of multiple mathematical theories. Say we have two math theories p that both explain a desired set of theorems X and avoid an undesired set of theorems Y, and we want to guess if a new theorem z is true or not. I.e.,

$$X \subset U(p), \ U(p) \cap Y = \emptyset, \ z \in U(p) \ ?$$

Suppose that our first theory p is three bits long, our second theory is five bits long, and that the first includes z while the second doesn't. Then as before we get the respective weights

$$\frac{1/8}{1/8 + 1/32}, \frac{1/32}{1/8 + 1/32}$$

in favor of theorem z and against theorem z.

---

[26] At first Solomonoff had some trouble making this method converge; he was not aware of the crucial idea of self-delimiting programs, which makes it easy to sum over all possible programs for something. That algorithmic information theory should be based on self-delimiting programs was independently realized in the 1970s by L. A. Levin and Chaitin.

These are *a priori* estimates based only on considering conceptual complexity, on the fact that we believe in theories because they are compressions. Ideally, we should sum over *all* theories that match what we already know to make our predictions, but this cannot be done in practice. We can only make rough estimates of this ideal prediction.

Examples of randomness in real mathematics

After Turing discovered the halting problem, there was a lot of work that revealed that the halting problem occurs in many fields of mathematics. This work was unfortunately never collected in a single place, it was never put together in a definitive treatise. Nevertheless we can use it to find the halting probability in many fields of pure mathematics.

In quantum physics you can predict probabilities, not individual events. In pure math, if you can show that something encodes the bits of the halting probability, then you immediately get very precise statistical information about it, even though individual cases are impossible to determine. And the halting problem and halting probability are ubiquitous in pure mathematics, because *universal computation is ubiquitous.* You can easily build a universal computer using combinatorial components from just about any field of discrete pure math.[27] And as we will show in the next two chapters, this can also be done with continuous mathematics, with classical questions in mathematical analysis and mathematical physics.

The fundamental question then becomes, how natural are these examples of uncomputability and even algorithmic irreducibility and randomness in many different fields of mathematics? Are these examples artificial, degenerate cases, or are they natural and symptomatic? With elegant programs, we saw a situation in which only finitely many truths of a certain kind are provable. When we study computable reals later in this chapter, we will see a situation in which the difficulties are pervasive. Most reals are uncomputable, with probability one. There are of course many computable reals, but they nevertheless have total probability zero.

Because of examples like these we believe that uncomputability, undecidability and incompleteness are pervasive, that these are not isolated phenomena. And we are doing our best to also convince you of this!

Now we're going to show in some detail that the algorithmic randomness and irreducibility of the bits of Ω is also found in real mathematics, namely in number theory and algebra. Let's start with number theory.

Universal Diophantine equations

A Diophantine equation is an equation involving only whole numbers.

---

[27]See Stephen Wolfram, *A New Kind of Science* and Chaim Goodman-Strauss "Can't Decide? Undecide!", *AMS Notices*, March 2010, pp. 343–356.

There is a single polynomial Diophantine equation which is a universal Turing machine:

$$P_U(i, k, n, x_1, x_2, \ldots, x_m) = 0$$

has a solution with non-negative integers $x_1, \ldots, x_m$ iff the universal Turing machine eventually halts when started with program i and inputs k and n. Actually, if the program i halts on inputs k and n, then P_U has *infinitely many solutions*, and if i fails to halt on inputs k and n, then P_U has no solutions.

Instead of an infinite number of solutions, we would prefer there to be a single solution if i halts on inputs k and n. This can be achieved by using an exponential Diophantine equation instead of a polynomial Diophantine equation.

An exponential Diophantine equation is a Diophantine equation in which we allow x^y as well as x^3, i.e., variables can also appear in exponents, which is not the case in polynomial Diophantine equations, where exponents must always be constants.

There is a single exponential Diophantine equation which is a universal Turing machine:

$$E_U(i, k, n, x_1, x_2, \ldots, x_m) = 0$$

has a solution with non-negative integers $x_1, \ldots, x_m$ iff the universal Turing machine eventually halts when started with program i and inputs k and n. Furthermore if the program i halts on inputs k and n, then E_U has *exactly one solution*, and if i fails to halt on inputs k and n, then E_U has no solutions.

Let's approximate Ω, more precisely, let's get better and better lower bounds on Ω. Consider this computable sum:

$$\Omega_n = \sum_{|p| \leq n \ \& \ U(p) \text{ halts in time } \leq n} 2^{-|p|}. \tag{18}$$

Ω_n is computable and tends to Ω in the limit from below:

$$\Omega_1 \leq \Omega_2 \leq \Omega_3 \ldots \to \Omega. \tag{19}$$

This converges to Ω, but this happens immensely slowly, and there is no way to know how far out to go to get a given degree of accuracy.

To use our universal Diophantine equations P_U and E_U, we consider a program i that computes the kth bit of Ω_n and then loops forever if this bit is a 0 and halts if this bit is a 1. Note that since $\Omega_n \to \Omega$, for all sufficiently large n the kth bit of Ω_n will be correct.[28] Plug into our universal Diophantine equations the program i. This turns P_U into P_Ω and E_U into E_Ω. More precisely, this choice of i makes

$$P_U(i, k, n, x_1, x_2, \ldots, x_m) = 0 \tag{20}$$

---

[28]This works because Ω must be irrational, but one can avoid this issue by stipulating that the base-two representation for Ω must have infinitely many 1s. In other words, if there are only finitely many 1s, take the last 1 in Ω, which comes before an infinite list of 0s, and replace it by a 0 followed by an infinite list of 1s.

into

$$P_\Omega(k, n, x_1, x_2, \ldots, x_m) = 0, \tag{21}$$

and makes

$$E_U(i, k, n, x_1, x_2, \ldots, x_m) = 0 \tag{22}$$

into

$$E_\Omega(k, n, x_1, x_2, \ldots, x_m) = 0. \tag{23}$$

Now let's ask:

Does a Diophantine equation have finitely or infinitely many solutions?

In 1900, David Hilbert asked for a general method to determine whether or not a Diophantine equation has a solution. In 1970 Yuri Matiyasevich showed that there is no general method, since this is equivalent to solving the halting problem. He did this by constructing the universal Diophantine equation given above. In 1984 he and James Jones came up with a particularly elegant construction for a universal exponential diophantine equation. We can do even better: Instead of just getting the halting problem, we can get the halting probability by using an individual Diophantine equation.

Here is how you get the bits of Ω from P_Ω. If we fix k, the number of n such that

$$P_\Omega(k, n, x_1, x_2, \ldots, x_m) = 0 \tag{24}$$

has a solution will be finite or infinite depending on whether the kth bit of Ω is, respectively, a 0 or a 1.

And here is how you get the bits of Ω from E_Ω. If we fix k, the number of solutions of

$$E_\Omega(k, n, x_1, x_2, \ldots, x_m) = 0 \tag{25}$$

will be finite or infinite depending on whether the kth bit of Ω is, respectively, a 0 or a 1.[29]

Does a Diophantine equation have an even/odd number of solutions?

Following Ord and Kieu (2003), we can change P_Ω to another polynomial P'_Ω with the following property: If we fix k, the number of n such that

$$P'_\Omega(k, n, x_1, x_2, \ldots, x_m) = 0 \tag{26}$$

has a solution will always be finite and it will be even or odd depending on whether the kth bit of Ω is, respectively, a 0 or a 1. To see how to produce P'_Ω, which is more complicated than P_Ω, see Chaitin, *Meta Math!*.

---

[29]In Chaitin, *Algorithmic Information Theory*, 1987, this equation is worked out in detail using LISP and the method of Matijasevic and Jones, which is based on a beautiful lemma of Édouard Lucas that the coefficient of x^k in the expansion of $(1 + x)^n$ is odd iff each bit in k bitwise implies the corresponding bit in n. The result is a 200-page exponential diophantine equation in 20,000 unknowns, parts of which are printed in *Algorithmic Information Theory*.

Similarly, we can change $E_\Omega = 0$ into another exponential Diophantine equation $E'_\Omega = 0$ with the following property: If we fix k, the number of solutions of

$$E'_\Omega(k, n, x_1, x_2, \ldots, x_m) = 0 \qquad (27)$$

will always be finite and it will be even or odd depending on whether the kth bit of Ω is, respectively, a 0 or a 1.

The word problem for semi-groups

Enough of number theory! Now for some algebra. Given a finite set of generators (letters), e.g., a, b, c, and a finite set of relations between them, e.g.,

$$aab = ca, \; ab = ba, \; acb = c, \qquad (28)$$

are two words equivalent? Questions of this kind were originally studied by the Norwegian mathematician Axel Thue.

Following Emil Post (1947), we can represent what is called the *instantaneous description* of a Turing machine by a word bounded on both sides by the special symbol λ, and giving the current contents of the Turing machine tape with a symbol denoting the current state q inserted just to the left of the current position of the read/write head. Then for each Turing machine there is a finite set of relations over a finite alphabet such that

$$\lambda q_{\text{initial}} xyz \ldots \lambda = \lambda q_{\text{final}} \lambda \qquad (29)$$

if and only if the Turing machine eventually halts when started with $xyz \ldots$ on the tape and with the read/write head scanning x.

To get each of the bits of Ω we pick a specific Turing machine very carefully,[30] we convert it into a word problem, and then we ask the following questions. For each k, are there finitely or infinitely many n such that

$$\lambda q_{\text{initial}} a^k b^n \lambda = \lambda q_{\text{final}} \lambda \; ? \qquad (30)$$

The answer determines the kth bit of Ω. Alternatively, for a different choice of alphabet and relations, again for each k, are there an even or an odd number of n such that this equality holds? If the alphabet and finite set of relations are chosen properly, the answer again determines the kth bit of Ω.[31]

So Ω's irreducible complexity can be found in many (most?) fields of pure mathematics. God plays dice everywhere in pure math![32]

---

[30] As before, this Turing machine computes the kth bit of Ω_n and then loops forever if this bit is a 0 and halts if this bit is a 1.

[31] For an extremely clear explanation of the method used by Post in 1947, see the article by Martin Davis "What is a Computation?" in Cristian Calude, *Randomness and Complexity, from Leibniz to Chaitin*.

[32] For a work of fiction reacting to this, see Arturo Sangalli, *Pythagoras' Revenge*, reviewed in the May 2010 issue of the *AMS Notices*.

How real are the real numbers? Borel 1927, 1952 and Turing 1936 revisited

Now we'd like to consider a path leading to Ω that makes clearer how it works, why Ω has the properties that it does, and provides some historical context for appreciating Ω. We'll also consider a domain in which the solvable problems have probability zero and the unsolvable problems have probability one, which is even worse than what happens with elegant programs, for which only finitely many are provably elegant and infinitely many are unprovably elegant.

Turing, 1936: There are more uncomputable reals than computable reals

It is not usually remembered that Turing's famous 1936 paper "On Computable Numbers…" deals with computable and uncomputable infinite-precision real numbers, something which one never sees in a modern computer, where all reals are finite precision. A real is computable if there is an un-ending algorithm that will compute it digit by digit with arbitrarily high precision, and a real is uncomputable if this is impossible to do.

Turing immediately notes that the computable reals are only as numerous as the set of possible computer programs, which is only a countable set, while the uncomputable reals are just as numerous as the set of all reals. So:

$$\#\{\text{computable reals}\} = \aleph_0, \quad \#\{\text{uncomputable reals}\} = 2^{\aleph_0}.$$

Uncomputable reals have probability one, computable reals have probability zero

In fact, a simple measure-theoretic or probabilistic argument shows that if you pick a real x at random between 0 and 1, it is possible but infinitely unlikely to be computable.

In other words, consider a real $x \in [0,1] \equiv \{x : 0 \le x \le 1\}$ with the uniform probability distribution. Then

$$\textbf{Prob}\{\text{computable reals}\} = 0, \quad \textbf{Prob}\{\text{uncomputable reals}\} = 1.$$

Here is a proof that the computable reals have measure zero. Cover the first computable real with an interval of size $\epsilon/2$, the second computable real with an interval of size $\epsilon/4$, etc. The total size of the covering is

$$\le \frac{\epsilon}{2} + \frac{\epsilon}{4} + \frac{\epsilon}{8} + \frac{\epsilon}{16} + \frac{\epsilon}{32} + \cdots = \epsilon,$$

which we can make as small as desired. Hence the computable reals are a set of measure zero and have zero probability, and the uncomputable reals have probability one.

This proof uses ideas that were pioneered by Borel. But Borel never states this result. Instead he proves a much stronger, a much stranger result:

Borel 1952: Un-nameable reals have probability one

(And nameable reals have probability zero.) In his last book, published when he was in his 80s, Émile Borel points out that most real numbers cannot even be individually named, constructively or non-constructively:

Prob{nameable reals} = 0, **Prob**{un-nameable reals} = 1.

This follows immediately just as before, since the set of all possible names for reals is a subset of the set of all possible strings over a finite alphabet, and is therefore countable, just like the set of all possible computer programs and the set of all possible algorithms for computing a computable real.

Borel's 1927 oracle number: Nth bit answers the Nth yes/no question

Borel felt more comfortable with countable sets than with uncountable sets. He also preferred computable functions to uncomputable functions. In 1927 he gave a wonderful example of how unreal a real number could be: his amazing know-it-all oracle number written in base-two whose Nth bit after the decimal point answers the Nth yes/no question. Here we take advantage of the fact that the set of all possible yes/no questions is included in the set of all possible strings over a finite alphabet, and is therefore countable.

Unfortunately, Borel's oracle real is hard to define rigorously. For example, consider the following question:

Is the answer to this question "no"?

There is no valid answer.

Or we can ask about our future behavior and then do the opposite.

However, just number all the possible computer programs. Then it is easy to make a rigorous version of the Borel oracle number that answers all instances of Turing's halting problem:

"Borel-Turing" oracle number: Nth bit tells us if the Nth program halts

Unfortunately this Borel-Turing oracle number — by the way, Borel and Turing seemed to be totally unaware of each other's work — is extremely redundant; it repeats a lot of information.

Why?

Because N instances of the halting problem is only equal to $\log_2 N$ bits of mathematical information, not to N bits of mathematical information. The Borel-Turing number is highly redundant.

Here is a proof of this. Consider N bits of the Borel-Turing number, i.e., N cases of the halting problem. You can tell which of these N programs halt if you know *how many of them* halt. Just run the N programs in parallel until that many have halted. The remaining programs will never halt.

By using a slightly more sophisticated version of this idea, we finally arrive at the halting probability

$$\Omega \equiv \sum_{U(p) \text{ halts}} 2^{-|p|}, \tag{31}$$

which contains no redundancy and is the best possible way to pack the answers to every possible case of the halting problem in a single oracle real.

Here is how this works. It works because:

First N bits of Ω tell us which $\leq N$ bit programs halt

Why is this? Well, do you remember those lower bounds Ω_k on Ω that we considered earlier in this chapter:

$$\Omega_k = \sum_{|p| \leq k \ \& \ U(p) \text{ halts in time } \leq k} 2^{-|p|} \ ?$$

Ω_k is computable and tends to Ω in the limit from below:

$$\Omega_1 \leq \Omega_2 \leq \Omega_3 \ldots \to \Omega.$$

So if you are given a minimum-size program $\Omega^\star$ that calculates the first N bits of Ω, we can concatenate in front of $\Omega^\star$ a prefix π_Ω that does the following: First π_Ω runs $\Omega^\star$ to calculate the first N bits of Ω. Then π_Ω calculates better and better lower bounds on Ω, Ω_k, for $k = 1, 2, 3, \ldots$ until the first N bits of Ω_k are correct, i.e., the same as the first N bits of Ω.

This is possible because Ω must be irrational, but just in case we happen not to know this, just do all of this with a base-two representation for Ω in which there are infinitely many 1s. In other words, if there are only finitely many 1s, take the last 1, which comes before an infinite list of 0s, and replace it by a 0 followed by an infinite list of 1s.

Once we have found a value of k for which the first N bits of Ω_k are correct, we know every program that is $\leq N$ bits in size that ever halts (each of them halts in time $\leq k$), because if another $\leq N$ bit program were then to halt, the value of Ω_k would become greater than Ω, which contradicts the fact that Ω_k is always $\leq \Omega$.

Once we have found all $\leq N$ bit programs that halt, we can run them all to see what they produce, and then pick as our very own output the first positive integer M that does not have $H(M) \leq N$. So finally π_Ω produces as its output the first positive integer M with $H(M) > N$.

In summary,

$$U(\pi_\Omega \Omega^\star) = M \tag{32}$$

and

$$H(M) > N, \tag{33}$$

and so

$$|\pi_\Omega \Omega^\star| = c + H(\text{the first } N \text{ bits of } \Omega) > N \qquad (34)$$

and

$$H(\text{the first } N \text{ bits of } \Omega) > N - c, \qquad (35)$$

which was to be proved.

$H(\text{First } N \text{ bits of } \Omega) > N - c$, and Ω is irreducible

This shows that unlike the Borel-Turing oracle number, Ω is an oracle for the halting problem that is incompressible. From this fact, using the ideas in our proof that it takes an N-bit theory to prove that an N-bit program is elegant, it is not difficult to show that it takes an N-bit theory to determine N bits of Ω, which is the *logical irreducibility* of Ω, and has a much greater epistemological impact than the mere fact that Ω is algorithmically irreducible and algorithmically random.

On the other hand, it is the fact that Ω is algorithmically irreducible that makes it look random, accidental, undistinguished, typical, even though it is jam-packed with useful information about the halting problem. For if Ω had any redundancy, then it would not be the best possible oracle for the halting problem. Any time we eliminate all the redundancy from something, we get randomness! And that's why pure math is full of randomness.

And most real numbers are rather unreal. Ω is a violently unreal real number that we have tried to make as real as possible. In other words, Ω is a specific real that lies just on the border between the computable and the violently uncomputable. As the series of approximations Ω_k show, Ω is almost real, but not quite. And that is why it is so very interesting. Reals are un-nameable with probability one, but we can never exhibit a specific example, because then we would actually be naming such a real. But in the case of Ω, there it is, a specific, fairly natural example of algorithmic irreducibility and randomness.

Mathematics, biology and metabiology

We've discussed physical and mathematical theories in this chapter; now let's turn to biology, the most exciting field of science at this time, but one where mathematics is not very helpful. Biology is very different from physics. There is no simple equation for your spouse. Biology is the domain of the complex. There are not many universal rules. There are always exceptions. Math is very important in theoretical physics, but there is no fundamental mathematical theoretical biology.

This is unacceptable. The honor of mathematics requires us to come up with a mathematical theory of evolution and either prove that Darwin was wrong or right! We want a general, abstract theory of evolution, not an immensely complicated theory of actual biological evolution. And we want proofs, not computer simulations! So we've got to keep our model very, very simple.

That's why this proposed new field is *metabiology*, not biology.

What kind of math can we use to build such a theory? Well, it's certainly not going to be differential equations. Don't expect to find the secret of life in a differential equation; that's the wrong kind of mathematics for a fundamental theory of biology.

In fact a universal Turing machine has much more to do with biology than a differential equation does. A universal Turing machine is a very complicated new kind of object compared to what came previously, compared with the simple, elegant ideas in classical mathematics like analysis. And there are self-reproducing computer programs, which is an encouraging sign.

There are in fact three areas in our current mathematics that do have some fundamental connection with biology, that show promise for math to continue moving in a biological direction:

Computation, Information, Complexity.

DNA is essentially a programming language that computes the organism and its functioning; hence the relevance of the theory of computation for biology.

Furthermore, DNA contains biological information. Hence the relevance of information theory. There are in fact at least four different theories of information:

- Boltzmann statistical mechanics and Boltzmann entropy,

- Shannon communication theory and coding theory,

- algorithmic information theory (Solomonoff, Kolmogorov, Chaitin), which is the subject of this chapter, and

- quantum information theory and qubits.

Of the four, AIT (algorithmic information theory) is closest in spirit to biology. AIT studies the size in bits of the smallest program to compute something. And the complexity of a living organism can be roughly (very roughly) measured by the number of bases in its DNA, in the biological computer program for calculating it.

Finally, let's talk about complexity. Complexity is in fact the most distinguishing feature of biological as opposed to physical science and mathematics. There are many computational definitions of complexity, usually concerned with computation times, but again AIT, which concentrates on program size or conceptual complexity, is closest in spirit to biology.

Let's emphasize what we are not interested in doing. We are certainly not trying to do systems biology: large, complex realistic simulations of biological systems. And we are not interested in anything that is at all like Fisher-Wright population genetics that uses differential equations to study the shift of gene frequencies in response to selective pressures.

We want to use a sufficiently rich mathematical space to model the space of all possible designs for biological organisms, to model biological creativity. And the only space that is sufficiently rich to do that is a software space, the space of all possible algorithms in a fixed programming language. Otherwise we have limited ourselves to a fixed set of possible genes as in population genetics, and it is hopeless to expect to model the major transitions in biological evolution such as from single-celled to multicellular organisms, which is a bit like taking a main program and making it into a subroutine that is called many times.

Recall the cover of Stephen Gould's *Wonderful Life* on the Burgess shale and the Cambrian explosion? Around 250 primitive organisms with wildly differing body plans, looking very much like the combinatorial exploration of a software space. Note that there are no intermediate forms; small changes in software produce vast changes in output.

So to simplify matters and concentrate on the essentials, let's throw away the organism and just keep the DNA. Here is our proposal:

> *Metabiology: a field parallel to biology that studies the random evolution of artificial software (computer programs) rather than natural software (DNA), and that is sufficiently simple to permit rigorous proofs or at least heuristic arguments as convincing as those that are employed in theoretical physics.*

This analogy may seem a bit far-fetched. But recall that Darwin himself was inspired by the analogy between artificial selection by plant and animal breeders and natural section imposed by malthusian limitations.

Furthermore, there are many tantalizing analogies between DNA and large, old pieces of software. Remember *bricolage*, that Nature is a cobbler, a tinkerer? In fact, a human being is just a very large piece of software, one that is 3×10^9 bases $= 6 \times 10^9$ bits $\approx$ one gigabyte of software that has been patched and modified for more than a billion years: a tremendous mess, in fact, with bits and pieces of fish and amphibian design mixed in with that for a mammal.[33] For example, at one point in gestation the human embryo has gills. As time goes by, large human software projects also turn into a tremendous mess with many old bits and pieces.

The key point is that you can't start over, you've got to make do with what you have as best you can. If we could design a human being from scratch we could do a much better job. But we can't start over. Evolution only makes small changes, incremental patches, to adapt the existing code to new environments.

So how do we model this? Well, the key ideas are:

Evolution of mutating software,

and:

Random walks in software space.

That's the general idea. And here are the specifics of our current model, which is quite tentative.

---

[33]See Neil Shubin, *Your Inner Fish: A Journey into the 3.5-Billion-Year History of the Human Body.*

We take an organism, a single organism, and perform random mutations on it until we get a fitter organism. That replaces the original organism, and then we continue as before. The result is a random walk in software space with increasing fitness, a hill-climbing algorithm in fact.[34]

Finally, a key element in our proposed model is the definition of fitness. For evolution to work, it is important to keep our organisms from stagnating. It is important to give them something challenging to do.

The simplest possible challenge to force our organisms to evolve is what is called the Busy Beaver problem, which is the problem of providing concise names for extremely large integers. Each of our organisms produces a single positive integer. The larger the integer, the fitter the organism.[35]

The Busy Beaver function of N, BB(N), that is used in AIT is defined to be the largest positive integer that is produced by a program that is less than or equal to N bits in size. BB(N) grows faster than any computable function of N and is closely related to Turing's famous halting problem, because if BB(N) were computable, the halting problem would be solvable.[36]

Doing well on the Busy Beaver problem can utilize an unlimited amount of mathematical creativity. For example, we can start with addition, then invent multiplication, then exponentiation, then hyper-exponentials, and use this to concisely name large integers:

$$ N + N \;\rightarrow\; N \times N \;\rightarrow\; N^N \;\rightarrow\; N^{N^N} \;\rightarrow\; \ldots $$

There are many possible choices for such an evolving software model: You can vary the computer programming language and therefore the software space, you can change the mutation model, and eventually you could also change the fitness measure. For a particular choice of language and probability distribution of mutations, and keeping the current fitness function, it is possible to show that in time of the order of 2^N the fitness will grow as BB(N), which grows faster than any computable function of N and shows that genuine creativity is taking place, for mechanically changing the organism can only yield fitness that grows as a computable function.[37]

So with random mutations and just a single organism we actually do get evolution, unbounded evolution, which was precisely the goal of metabiology!

---

[34]In order to avoid getting stuck on a local maximum, in order to keep evolution from stopping, we stipulate that there is a non-zero probability to go from any organism to any other organism, and $-\log_2$ of the probability of mutating from A to B defines an important concept, the *mutation distance*, which is measured in bits.

[35]*Alternative formulations:* The organism calculates a total function $f(n)$ of a single non-negative integer n and $f(n)$ is fitter than $g(n)$ if $f(n)/g(n) \rightarrow \infty$ as $n \rightarrow \infty$. Or the organism calculates a (constructive) Cantor ordinal number and the larger the ordinal, the fitter the organism.

[36]Consider BB$'(N)$ defined to be the maximum run-time of any program that halts that is less than or equal to N bits in size.

[37]Note that to actually simulate our model an oracle for the halting problem would have to be employed to avoid organisms that have no fitness because they never calculate a positive integer. This also explains how the fitness can grow faster than any computable function. In our evolution model, implicit use is being made of an oracle for the halting problem, which answers questions whose answers cannot be computed by any algorithmic process.

This theorem may seem encouraging, but it actually has a serious problem. The times involved are so large that our search process is essentially *ergodic*, which means that we are doing an exhaustive search. Real evolution is not at all ergodic, since the space of all possible designs is much too immense for exhaustive search.

It turns out that with this same model there is actually a much quicker *ideal evolutionary pathway* that achieves fitness BB(N) in time of the order of N. This path is however unstable under random mutations, plus it is much too good: Each organism adds only a single bit to the preceding organism, and immediately achieves near optimal fitness for an organism of its size, which doesn't seem to at all reflect the haphazard, frozen-accident nature of what actually happens in biological evolution.[38]

So that is the current state of metabiology: a field with some promise, but not much actual content at the present time. The particular details of our current model are not too important. Some kind of mutating software model should work, should exhibit some kind of basic biological features. The challenge is to identify such a model, to characterize its behavior statistically,[39] and *to prove* that it does what is required.

Post Scriptum

After this chapter was completed the mathematical structure of metabiology fell into place. The crucial step was to permit *algorithmic* mutations: If a mutation M is a K-bit program that takes the original organism A as input and produces the mutated organism $A' = M(A)$ as output, then this mutation M has probability 2^{-K}. Then it becomes possible to show that evolution rapidly takes place in the Busy Beaver model of evolution discussed above. In fact, better and better lower bounds on the halting probability Ω will rapidly evolve.

For the mathematical details, see G. J. Chaitin, "Life as evolving software," to be published in H. Zenil, *A Computable Universe*, World Scientific, 2012. For a non-technical book-length treatment, see G. Chaitin, *Proving Darwin: Making Biology Mathematical* to be published by Pantheon in 2012. For an overview of this book, a lecture entitled "Life as evolving software," go to *www.youtube.com* and search for *chaitin ufrgs*.

---

[38]The Nth organism in this ideal evolutionary pathway is essentially just the first N bits of the numerical value of the halting probability Ω. Can you figure out how to compute BB(N) from this?

[39]For instance, will some kind of hierarchical structure emerge? Large human software projects are always written that way.

3. A List of Problems

THE LATE BRAZILIAN FUNCTIONAL ANALYST LEOPOLDO NACHBIN used to say that Rio's scenery is inspiring for mathematicians. He lived for a while in an apartment by the beach in Ipanema and would take his friends to the large panoramic window that opened over the sands sung in *The Girl from Ipanema* and would say, from such a beautiful view beautiful theorems will be born. (Nachbin is also remembered because of a witty answer he gave to a journalist who was interviewing him. Nachbin said: a human science like mathematics... The interviewer jumped: but mathematics isn't one of the human sciences! And Nachbin had the last word: how's that possible? Then who does mathematics?)

Nachbin was Doria's PhD advisor, and when his former student approached him with a few questions on the foundations of physics (as Doria had been trained as an engineer with graduate degrees in theoretical physics), Nachbin immediately introduced Doria to Newton da Costa.

The questions Doria asked Nachbin were:

- Does Cantor's Continuum Hypothesis matter for theoretical physics?

- Are there actual physical processes that can settle undecidable sentences in some formal system? Or processes which can compute uncomputable stuff?

Mr. Contradiction: Newton da Costa

We start with a small vignette written by one of Newton da Costa's coworkers:

> Newton da Costa is a suave, dignified looking gentleman who frequently dons a slightly ironic smile on his face. His blond, thinning hair is always carefully combed and his glasses increase the size of his always alert eyes. He is eighty now[40] but looks in his late sixties.
>
> When in his teens he asked himself: why can't we deal with contradictions in mathematics? For contradictions exist everywhere in the real world. Witnesses give conflicting depositions in court; when a doctor examines a patient during a medical consultation in search for the causes of some ailment, the data obtained are frequently conflicting data.

Why are contradictions excluded from mathematical arguments?

Other mathematicians and logicians had asked the same question in the early 20th century, such as the Russian Vasiliev (who developed a system of "imaginary logics") and the Pole Jaśkowski, who invented a propositional

---

[40]He was born in 1929.

calculus that allows for contradictions. Newton da Costa did his work independently of both Vasiliev and Jaśkowsky, and created a full system of propositional and predicate logics with the possibility of contradictions.

(Predicate logics have predicate symbols like $P(x)$, which may be read as "x has the property P," and it is known that classical predicate logic plus the Zermelo–Fraenkel axioms are enough the describe most of the work so far done by mathematicians — we will elaborate later on the stuff that remains ouside the grasp of ZF's axioms.)

da Costa developed his "paraconsistent systems" in a series of papers published mainly in Euriopean journals in the 1960s; they are now widely applied in several domains, from the development of experts systems that deal with conflicts to theoretical constructions in mathematics — a very interesting result is the proof by economist Fernando Garcia that Arrow's Impossibility Theorem in economics cannot be proved in several paraconsistent logic systems.

An aside: on paraconsistent logics

Newton da Costa is best known for his contributions to the field of nonclassical logics. His paraconsistent logics are examined in a long quote from the introduction of the chapter he wrote (with Décio Krause and Otávio Bueno) on his logics for the *Handbook of the Philosophy of Science* in its vol. 5, *Philosophy of Logic*[41]:

> The origins of paraconsistent logics go back to the first systematic studies dealing with the possibility of rejecting or restricting the law (or principle) of non-contradiction, which (in one of its possible formulations) says that a formula and its negation cannot both be true. The law of non-contradiction is one of the basic laws of traditional, or classical (Aristotelian), logic. This principle is important. After all, since inconsistency entails triviality in classical systems, an inconsistent set of premises yields any well-formed statement as a consequence. The result is that the set of consequences of an inconsistent theory, or set of premises, will explode into triviality and the theory is rendered useless.
>
> Another way of expressing this fact is by saying that under classical logic the closure of any inconsistent set of sentences includes every sentence. It is this which lies behind Sir Karl Popper's famous statement that the acceptance of inconsistency "... would mean the complete breakdown of science" and that an inconsistent system is ultimately uninformative.
>
> Inconsistencies appear in various levels of discussion of science and philosophy. For instance, Charles Sanders Peirce's world of 'signs' (in which we inhabit) is an inconsistent and incomplete world. Niels Bohr's theory of the atom is one of the well-known examples in science of an inconsistent theory. The old quantum theory of black-body radiation, Newtonian cosmology, the (early) theory of infinitesimals in the calculus, the Dirac

---

[41]We have slightly retouched the text in order to make it less technical-sounding.

δ-function, Stokes' analysis of pendulum motion, Michelson's 'single-ray' analysis of the Michelson–Morley interferometer arrangement, among others, can also be considered as cases of inconsistencies in science. Given cases such as these, it seems clear that we should not eliminate a priori inconsistent theories, but rather investigate them. In this context, paraconsistent logics acquire a fundamental role within science itself as well as in its philosophy. As we will see below, due to the wide range of applications which nowadays have been found for these logics, they have an important role in applied science as well.

The forerunners of paraconsistent logics are Jan Łukasiewicz and Nikolai I. Vasiliev. Independently of each other, both suggested in 1910 and 1911 that 'non-Aristotelian' logics could be obtained by rejecting the law of non-contradiction. Although Łukasiewicz did not construct any system of paraconsistent logic, his ideas on the principle of non-contradiction in Aristotle influenced his student S. Jaśkowski in the construction of 'discussive' (or 'discursive') logic in 1948. (We will comment on Jaśkowski's systems below.) In 1911, 1912 and 1913, inspired by the works of Lobachewski on non-Euclidean geometry, initially called 'imaginary geometry,' Vasiliev envisaged an imaginary logic, a non-Aristotelian logic where the principle of non-contradiction was not valid in general. According to the late logician Ayda Arruda, Vasiliev did not believe that there exist contradictions in the real world, but only in a possible world created by the human mind. Thus, he hypothesized imaginary worlds where the Aristotelian principles were not valid, even though Vasiliev did not develop his ideas in full.

The very first logician to construct a formal system of paraconsistent logic was Stanislaw Jaśkowski in 1948. His motivations came from his interests in systematizing theories that contain contradictions, such as dialectics, as well as to study theories where contradictions are caused by vagueness. He was also interested in the study of empirical theories whose postulates include contradictory assumptions. Despite the wide range of possible applications, Jaśkowski's discussive logic was restricted to the propositional level. In 1958, Newton da Costa, independently of Jaśkowski, began the general study of contradictory systems. Ever since, da Costa has developed several systems related to paraconsistency (for instance, 'paraclassical logic'), showing how to deal with inconsistencies from different perspectives. He apparently became the first logician to develop strong logical systems involving contradictions which could be useful for substantive parts of mathematics as well as the empirical and human sciences. It should be remarked that the adjective 'paraconsistent' (which means something like 'at the side of consistency') was suggested by Francisco Miró–Quesada, in 1976, in a letter to da Costa.

Already in the sixties, the interest in logics dealing with inconsistencies began in other parts of the world as well, particularly in Poland, Australia, United States, Italy, Argentina, Belgium, Ecuador, and Peru, mainly for its relations to da Costa's logics and to relevant and dialectical logic. Of course, in this paper, we cannot refer to all of these tendencies nor do justice to all the authors involved. For the historical details, we recommend the reading of the papers mentioned above.

At least two facts have contributed to emphasize the relevance of these developments. The first is that in 1990, Mathematical Reviews added a new entry, 03B53, called 'Paraconsistent Logic'. From 2000 on, the title was changed to 'Logics admitting inconsistency (paraconsistent logics, discussive logics, etc.)', thus encompassing a wider subject. The second fact is that since 1996, several World Congresses on Paraconsistency have been organized.[42] Nowadays 'paraconsistency' can be regarded as a field of knowledge. But perhaps the most surprising fact concerning paraconsistent logic is related to its applications. As we will note later, there have been applications not only to the foundations of science and its philosophical analysis, but even to technology. Here we do not have space to present all the details, but the references list the original sources.

[...]

We don't think there is just one 'true logic'. After all, distinct logical systems can be useful to describe different aspects of knowledge. (The same point can be made about distinct mathematical systems, and perhaps even about different physical systems.) In other words, we defend a form of logical pluralism. But our proposal is not relativist, since it's not the case that anything goes as far as applied logic is concerned. We can always rule out certain applied logical systems as being inadequate for certain domains. For instance, to capture constructive features of mathematical reasoning, classical logic is clearly inadequate; intuitionistic logic isn't. With regard to paraconsistent logics, we claim, with French philosopher Gilles Gaston Granger, that paraconsistent logic can, and should, be employed in the development of certain domains, but only as a preliminary tool. In the end, classical logic may eventually replace it as the underlying logic of these domains. Our position does not excludes Granger's.

In summary, there are in principle various 'pure' logics whose potential applications depend not only on a priori reasons, but, above all, on the nature of the applications one has in mind. This is also true of paraconsistent logic.[43]

How about undecidability and incompleteness? Yes, the original logical systems developed by da Costa exhibit the Gödel phenomenon, and are in that respect akin to their classical counterpart. Only the concept of consistency is changed to *non-triviality*: a formal system is trivial whenever each formal sentence in it is a theorem of the theory. That is to say, one proves everything,

---

[42]See www.cle.unicamp.br/wcp3 for the page of the third congress.
[43]We quote from [58].

and as a result the system is of non use. Then one shows that if the paraconsistent system isn't trivial, then one cannot prove a formal sentence that asserts it within the given paraconsistent system itself.

Now, back to our main theme.

A list of outlandish problems

Doria communicated to da Costa the questions he had asked his former advisor, and da Costa suggested that Doria should draw up a list of more specific questions that might lead do undecidability and incompleteness in sciences whose underlying language is mathematics.

The list deals with lots of delicate issues and became a kind of research program for da Costa and Doria. It includes:

1. *General relativity and Cohen's forcing.*

2. *The decision problem for chaotic systems.*

3. *Can we decide whether an equilibrium point in some system is stable or unstable?*

4. *Does Gödel incompleteness matter for the social sciences? For economics?*

5. *The Shannon theorems, entropy, communication systems, in different set-theoretic models.*

6. *The P vs. NP question.*

7. *Can we break the Turing barrier?*

Generic universes in gravitation theory

The idea here stems from Paul Cohen's forcing models for set theory that violate Cantor's Continuum Hypothesis. Given one of those models, pick up a line segment — a segment of the real numbers — that violate the Continuum Hypothesis. In Cohen's construction, they do so because there are new real numbers which are added to a previously obtained segment of reals in a Gödel constructive model.

Now there is a correspondence between, say, such a segment and all fields over a domain in spacetime. So, Cohen-expanded models have "new" fields. The question is: are these fields physically different from those in the constructive universe? If so, are there new physical effects associated to those new fields?

The Einstein equations describe the gravitational field. Local solutions are patched up to characterize universe models, that is, space-time manifolds (models for the universe) whose geometry directly stems out of known solutions for the gravitational equations.

We can also proceed the other way round: given a particular differentiable manifold, we can examine the space of all metric tensors on it. That space is what is called a Fréchet manifold, a rather nasty structure. Research in the late sixties and early seventies by Jerrold Marsden, Judy Arms and others has shown that symmetric solutions for the Einstein equations belong to a first-category set in that manifold, that is, a very small, sparse set, and it is known that non-symmetric solutions may exhibit some kind of chaotic behavior.

So we are left with the following questions: can we identify "topologically generic spacetimes" with "chaotic spacetimes"? (At least but for some reasonable, small set, of spacetimes which are exceptional in some ecognizable sense, whatever that might be.)

A second question: can we identify "topologically generic spacetimes" with "set-theoretically generic spacetimes" in a convenient set-theoretic model? Is there some kind of physically meaningful property that characterizes the 'interesting' generic metric tensors over a fixed differentiable manifold? (By 'physically meaningful' we think of a property which explicitly involves only 'physical' concepts.)

Now we can ask the following question, which is in fact a conjecture:

> We conjecture that anything obtained in that direction will turn out to be undecidable in Zermelo–Fraenkel set theory and will remain undecidable even if we add to it the full axiom of choice. Moreover, we conjecture that the "typical" spacetime, in a reasonable characterization for topology and measure in the space of all spacetimes, will turn out to be exotic in the sense of differential topology, without cosmic time, that is, without a global time-arrow, and — if considered from the viewpoint of reasonable forcing models — set-theoretically generic.

Typical in the sense of measure means: with 100% probability. Typical in the sense of topology means the same thing, but without figures (there is a way we can distinguish in topology sparse, very rarefied sets, and big, large, thick, typical sets: rarefied sets are first-category sets; sets of typical objects are second category sets. It's a qualitative way of saying the same thing that the probabilistic evaluation does, this time without figures.

But there is a caveat here: there are typical sets in the sense of topology that are zero-probability sets in the sense of measure (probability), and vice-versa, for the two concepts do not always coincide. That kind of dissonant behavior makes things really interesting, by the way...

The last portion of the conjecture has to do with Big Bang models: can we have Cohen-generic Big Bang spacetimes? Which is their meaning? Which is the relation, if any, between Cohen-generic spacetimes and the so-called exotic four-dimensional manifolds?

Be patient: we will soon elaborate on those things.

Is stability decidable? Is chaos decidable?

This question goes back to 1963 when Edward Lorentz published his paper "Deterministic non-periodic flow" where he identifies through statistical tests a possibly chaotic behavior in deterministic differential equations. Soon other possible examples of the same kind of behavior were discovered and examined, and in 1966 Steve Smale invented his famous horseshoe attractor and showed that it was associated to chaotic behavior in dynamical systems.

One then had to marry both sides of the study of chaotic systems: on one side we had several interesting examples of phenomena that originated in ireal-life situations and which seemed to exhibit chaotic behavior. On the other hand we had several theoretical constructs such as horseshoes or strange attractors that led to proved chaotic behavior.

Could we join the two pieces of the puzzle?

That problem was originally formulated by Morris Hirsch in 1983:

> *An interesting example of chaos—in several senses—is provided by the celebrated* Lorenz System. *[...] This is an extreme simplification of a system arising in hydrodynamics. By computer simulation Lorenz found that trajectories seem to wander back and forth between two particular stationary states, in a random, unpredictable way. Trajectories which start out very close together eventually diverge, with no relationship between long run behaviors.*

> *But this type of chaotic behavior has not been* proved. *As far as I am aware, practically nothing has been proved about this particular system. Guckenheimer and Williams proved that there do indeed exist many systems which exhibit this kind of dynamics, in a rigorous sense; but it has not been proved that Lorenz's system is one of them. It is of no particular importance to answer this question; but the lack of an answer is a sharp challenge to dynamicists, and considering the attention paid to this system, it is something of a scandal.*

> *The Lorenz system is an example of (unverified)* chaotic *dynamics; most trajectories do not tend to stationary or periodic orbits, and this feature is persistent under small perturbations. Such systems abound in models of hydrodynamics, mechanics and many biological systems. On the other hand experience (and some theorems) show that many interesting systems can be expected to be nonchaotic: most chemical reactions go to completion; most ecological systems do not oscillate unpredictably; the solar system behaves fairly regularly. In purely mathematical systems we expect heat equations to have convergent solutions, and similarly for a single hyperbolic conservation law, a single reaction-diffusion equation, or a gradient vectorfield.*

> *A major challenge to mathematicians is to determine which dynamical systems are chaotic or not. Ideally one should be able to tell from the form of the differential equations. The Lorenz system illustrates how difficult this can be.*

Smale formulated a particular version of that question: is Lorenz chaos due to the the Guckenheimer attractor? (A kind of horseshoe attractor.)

We will soon see how da Costa and Doria dealt with that problem.

Stable or unstable?

David Hilbert presented his list of 23 problems for 20th century mathematicians in 1900. In 1974 the American Mathematical Society invited several top-ranking mathematicians to a symposium where they discussed the status of the solutions to the Hilbert Problems and proposed new problems. Among those invited was Russian geometer Vladimir Arnold.

Arnold formulated a set of questions which are related to the chaos problem: can stability be algorithmically decided for polynomial dynamical systems over the integers? Such is the content of Arnold's 1974 Hilbert Symposium problems:

> Is the stability problem for stationary points algorithmically decidable? *The well-known Lyapounov theorem solves the problem in the absence of eigenvalues with zero real parts. In more complicated cases, where the stability depends on higher order terms in the Taylor series, there exists no algebraic criterion.*
>
> Let a vector field be given by polynomials of a fixed degree, with rational coefficients. Does an algorithm exist, allowing to decide, whether the stationary point is stable?
>
> *A similar problem: Does there exist an algorithm to decide, whether a plane polynomial vector field has a limit cycle?*

Arnold's queries have an intuitive meaning. Consider a marble in a bowl: if we slide it into the bowl it will spiral until it settles at the bottom of the bowl. Now invert the bowl and place the marble on the top: a small push will send it rushing down the external side of the bowl. In the first case we have a stable equilibrium; in the second case we have an unstable one.

In the general case is there an algorithm to decide whether, given the equations for the system, can we decide whether it will exhibit stable or unstable behavior?

The Hirsch question has been negatively answered by da Costa and Doria in 1991; Arnold's problems were also negatively solved in a more general setting (polynomials plus elementary functions, which can also be nicely coded in a Turing-machine tape) by them in 1993 and again negatively for singular isolated zeroes of polynomial dynamical systems over the integers in 1994.

Generic economies, generic social structures

Population dynamics is described by nonlinear differential equations. So, when we turn to the space of their solutions we can ask the same kind of questions we asked about general relativity. The questions on chaotic behavior turn out to be very difficult questions about the characterization of strange attractors and the like, but we can still ask about the overall relation between chaos and genericity, both topological and set-theoretic.

Now recall that John Nash proved in his PhD thesis in 1950 that every competitive game (a game without coalitions) has a kind of equilibrium solution we now call the Nash equilibrium. This result gave him the Nobel Prize more than forty years later, in 1994. Kenneth Arrow and Gérard Debreu used the Nash result to prove that every competitive market has equilibrium prices, and were duly rewarded with the 1972 Nobel in Economics.

Now the question is: the equilibria exist; can we compute them? Given sufficient computational power can we compute those equilibria beforehand? The idea is: since market stabilization may be a socially disruptive process, can we compute the equilibrium prices before an actual equilibrium is reached by the so-called "market forces"? That would spare us lots of trouble of course — if that's possible.

(That last question essentially arose out of a debate between Polish economist Oskar Lange and his Austrian colleague Ludwig von Mises.)

The question on population dynamics was dealt with in a 1994 paper by da Costa and Doria. The question about the computability of market equilibria in economics was suggested and solved by Brazilian economist MarceloTsuji in work co-authored by da Costa and Doria; it extends and slightly modifies previous results by Alain Lewis, and has a funny little story behind it, which we will tell in a moment. (We must recall that Vela Velupillai showed in the early 1990s that undecidability is a naturally occurring phenomenon in economic theory, so there isn't much of a surprise here.)

Beyond the Shannon theorems

The Shannon theorems are the essential tool in classical information theory. Claude Shannon had a degree in electrical engineering, and published in 1948 his famous essay on "A mathematical theory of communication," where he introduces what are now known as Shannon's theorems. The proof of these theorems, as offered by Shannon, is sketchy and full of gaps, and it took more than a decade of hard work by experts in probability theory to clean up everything. The key step is a deep result in measure and probability theory called Birkhoff's ergodic theorem, out of which the Shannon theorems smoothly follow.

The Shannon theorems impose a limit on the rate of transmission of information through a communication channel. There is a simple analogy at work here: think of a pipeline through which water flows. If it's of a small diameter less water will flow through it that if it has a large diameter. The pipeline becomes the communication channel in Shannon's construction, and its diameter the channel's capacity.

Water flow becomes information flow, and information is entropy, and old concept inherited from heat theory and from one of its underlying explanations, statistical mechanics. Entropy becomes information in 1928 in the work of Ralph Vinton Lyon Hartley (1888–1970), also from the Bell labs, which was essential to the development of the Shannon theorems.

There are however limits in the rate of information processing with controlled errors through a communication channel. That limit is given by the channel's capacity, which is expressed by an entropy related to the source and to the receiver. Now think of the following question: suppose that in a given set-theoretic model (or in any reasonable domain for the realization of information theory) we know that information cannot be trustworthily recovered in the receiver because the rate of transmission exceeds the channel's capacity; is there another model where, given the same 'syntactic' arrangement for that communication system, we will be able to recover any message from the source? (The idea is that we should somehow modify the Shannon inequalities when moving from one model to the other.)

Or, can we extend Shannon's information theory so that the extended theory allows the preceding construction?

The authors had already discussed the fact that entropy isn't clearly located within a set of trajectories in a chaotic dynamical system, e.g. in a Bernouilli shift, when we change the underlying set-theoretic model; however the example they give trivializes a communication system. Can we extend that to an useful violation of the Shannon inequalities? That question is still unanswered.

Does $P = NP$?

The last question in the updated 1994 list was already in the 1987 list and dealt with the P vs. NP conjecture, which has recently been the object of some published work by da Costa and Doria. They staunchily believe that $P = NP$ and its negation $P < NP$ will be found undecidable with respect to the full Zermelo–Fraenkel axiomatic system, even if one includes the axiom of choice in that system. Better: those sentences will remain undecidable even if we add to axiomatic set theory a string of large cardinal axioms.

This goes against the mainstream feeling in computer science, and we'll later present the case for such an unexpected behavior of the P vs. NP conjecture.

Their work on the P vs. NP question began in 1995 but was only published in early 2003. It remains as an alternative, still rather unexplored path. da Costa and Doria conjecture that $P < NP$ and $P = NP$ are independent of a whole hierarchy of axiomatic systems which have ZFC set theory at its bottom and large cardinals do not affect that independence. Wow! That's pretty technical! But just wait and see; everything will be explained in due course, as its starting point looks — looks, at least — quite intuitive.

da Costa and Doria also conjecture that a theory like Peano Arithmetic, which is the standard formalization for arithmetic, plus a reasonable infinitary proof rule such as Shoenfield's ω-rule, is enough to prove the Π_2 sentence $[P < NP]$. (Wait for the clarification!)

Hypercomputation

Another controversial and much-debated conjecture has to do with "breaking the Turing barrier." This is the subject of one of the sections in the last chapter in this book. da Costa and Doria have proved that there is an explicit expression for the halting function (the function that solves the halting problem, soon to be discussed) in the language of elementary analysis. One can use that function as an ideal analog computer, plug it into a Turing machine as an oracle and *voilà!* we have an ideal hypercomputer. More precisely, a machine that settles all arithmetical truths.

Can we do that in the real world?

Later on, as we will only explore the theories that try to go beyond the Church–Turing computation model in the last chapter of this book. The reason? It is outlandish and speculative, in good part. But we believe it will someday give rise to a working hypercomputer.

Outlandish stuff in general relativity

The initial exchanges between Doria and da Costa had to do with general relativity and gauge theory. Doria had been interested a few years before he met da Costa in the so-called gauge field copy problem. Let's talk a bit about it, and see why it leads to foundational issues in mathematics. Gauge fields are the main fare in the currently investigated unified theories in physics. They generalize electromagnetic fields, but as gauge fields are usually nonlinear stuff, they behave differently from electromagnetic fields in a number of situations. One of those is the gauge field copy phenomenon.

That phenomenon was only discovered by Nobel prizewinner Chen Ning Yang (together with his coworker Tai Tsun Wu) in 1975. Given a particular electromagnetic field, it is always derived from a unique potential, but for local gauge transformations (a gauge transformation is the way potentials transform both in electromagnetism and in gauge theories; they are local when we only consider them over restricted domains in spacetime). Here lies one of the main differences between the linear and nonlinear cases in gauge theories. For we may have the same nonlinear gauge field derived from two *physically different* potentials?

How can we say that they are different? For in one case the field is "source-less," and in another case it has a source. That means: in one case the field is — so to say — generated by the presence of matter, while on the other case it is a pure vacuum field. And one cannot make a gauge transformation that would map a nonzero matter source over a zero one, over empty space.

A final answer was reached in 1983; the copy phenomenon depended on some integrability conditions. And now for the metamathematical portion of it: if one looks at the space of all possible nonlinear gauge fields over some domain in spacetime, one sees that the fields that exhibit the copy phenomenon are arranged as a staircase-like structure called a stratification. That structure had been identified by Isadore Singer in 1978, and a bit later by Judy Arms and Jerrold Marsden in their study of the way the set of all gravitational fields (which can also be seen as a kind of gauge fiekds) organizes itself.

However each step in the ladder contained uncountably many fields. More precisely, the fields in each step of the ladder were as many as the real numbers in a line segment. So, Cohen's construction that led to a model disproving the continuum hypothesis might be applied here, who knows? Would that be possible? Would the new fields associated to Cohen's "generic" reals be physically different from the standard fields? Would Cohen's techniques matter for physics?

Paul Cohen and forcing

Paul Cohen (1934–2007) received his PhD at the University of Chicago with a thesis in analysis. It is said that he once had an exchange with a colleague that was doing work on logic and joked, there are no hard problems in logic, Cohen said. He immediately heard the reply: is that so? Then show the independence of Cantor's continuum hypothesis.

Cantor's continuum hypothesis has to do with orders of infinity. Cantor showed that there is a first kind of infinity: the number of all natural numbers $0, 1, 2, \ldots$. It is the countable infinity, noted $\aleph_0$ (aleph-zero or aleph-null). Then he showed that the number of real numbers between, say, two integers — take all reals between 0 and 1 — is strictly larger than $\aleph_0$, and can be written as an exponential:

$$2^{\aleph_0}.$$

Are there infinities between $\aleph_0$ and $2^{\aleph_0}$? Cantor believed that there are none, and formulated that belief as the Continuum Hypothesis (which we capitalize just once, to stress that we are here explaining its meaning).

(He called these numbers "transfinite numbers.")

Gödel proved the consistency of the continuum hypothesis.[44] He did so by his invention of the set-theoretic constructible universe, where each set is obtained out of a — possibly transfinite — sequence of operations that act on previously obtained sets. Very much like the way a building is constructed: the n-th floor can only be built when we have reached the $n - 1$-floor.

Cohen's idea was to introduce *generic reals* to increase the number of reals on a line segment, and therefore to be collected in a new set of intermediate transfinite size between the countable infinity and the cardinality of the reals. He did so by his totally original technique of forcing. Suppose that I wish to add a new real to the segment where one only has constructible reals. Our "new" real must:

- Be different from all previously existing reals.

- Have only a minimum of properties that would identify it as a real number, say, within a given interval.

---

[44] Actually Gödel proved the consistency of the Generalized Continuum Hypothesis.

Forcing was born out of the second condition. Cohen noticed that in order to specify some property of the reals one needed a finite number of bits.[45] Then we would only select the properties we required and "force" them to be true; its truthfulness would be ensured by some specific sequence of bits. They would then have to be combined in a consistent way. This procedure gives rise to an infinite sequence of bits that has to be shown to be different from all possible sequences that code already existing reals.

How would that be possible? Cohen availed himself of an apparent paradox that was discovered by logician Thoralf Skolem in the 1920s. Skolem noticed that there are countable models for set theory, that is, models for the Zermelo–Fraenkel (ZF) axioms that form a denumerable set (a set of cardinality $\aleph_0$) when seen from the outside, so to say. Why is that the case? Roughly and very informally because all properties that can be written in the language of ZF are a denumerable collection; a denumerable model simply lists all those properties and establish a correspondence between each property and the object that fulfills it.

So, if the model for ZF is denumerable, then so it any subset of the reals in it — when looked at from the outside. By a kind of diagonalization trick Cohen then shows that his "new" real is different from all previously existing reals in the original model.

The whole construction is a very delicate enterprise: one must first show that the extended collection (the model for ZF we start from plus the new real number, and its many consequences and objects that arise out of the new real, or of the mixup of old and new stuff) is in fact a model for axiomatic set theory. But it is. And follows Cohen's result.

Forcing is very abstract. A few years after Cohen obtained his result (which ensured him the Fields Medal in 1966) Dana Scott and Robert Solovay developed an alternative approach to Cohen-style independence proofs, known as the Boolean-valued models approach. Again the starting point is quite intuitive. We may represent a set A by its characteristic function f_A:

- Some x is in A if and only if $f_A(x) = 1$.

- Some x isn't in A if and only if $f_A(x) = 0$.

So it's indifferent to build a model with A or with f_A. Now suppose that we allow for a fuzzy situation, say, an element has 50% chances of being in some set. Then $f_A(x)$ would range in this case over values from 0% to 100%, or to avoid the percent notation, it would range over any value in the interval from 0 to 1. For technical reasons we can substitute the interval for an element of a Boolean algebra. The collection of such functions forms what we call a Boolean-valued model for set theory. The forcing proofs tend to be easier when done with the help of Boolean-valued models. And the interpretation of generic reals, and of generic objects, is easier to be figured out when one handles Boolean-valued models.

---

[45]Technically, this assumption is based on the so-called Compactness Theorem.

Now let's go back to general relativity and gauge theories. The idea went back to a suggestion — perhaps a tongue in cheek remart, at first — made by Paul Cohen in the 1966 article he wrote for *Scientific American* on his achievement. The title of the article was "Non-Cantorian Set Theory," and by that he meant models of set theory that violated Cantor's continuum hypothesis and the axiom of choice. He claims at the end of the article that one day non-Cantorian set theory will be as important for physics as are today the non-Euclidean geometries (e.g. in general relativity itself).

This seems to be the case, as in 1971 British mathematician Maitland-Wright noticed that in the so-called Solovay's mathematics (set theory with a weakened axiom of choice, among other things) every operator on Hilbert's space is bounded, a fact which may modify results in quantum mechanics, and may even be experimentally tested.

Why didn't da Costa and Doria go the way of quantum mechanics? Because there are constructions and proofs in quantum theory (e.g. the Feynmann integrals) whose mathematical foundations are unclear; and in order to apply forcing or Boolean models one requires a full consistent axiomatization of the theory we are dealing with.

General relativity and its axiomatics

The first step in applying forcing techniques to general relativity is to build an axiomatic framework for that theory, since forcing cannot be applied in intuitive settings. This was done with the help of an interpretation of the Suppes predicate technique developed a bit earlier by da Costa and Chuaqui.

But there is history behind it, for the axiomatization of physics goes back to Hilbert's 6th Problem, which is called the "ugly duckling" among the Hilbert Problems, as it has no possibly consensual solution. Here it is, in a translation of Hilbert's original formulation of the problem:

The Mathematical Treatment of the Axioms of Physics.

The investigations on the foundations of geometry suggest the problem: to treat in the same manner, by means of axioms, those physical sciences in which mathematics plays an important part; in the first rank are the theory of probability and mechanics.

As to the axioms of the theory of probabilities, it seems to me to be desirable that their logical investigation be accompanied by a rigorous and satisfactory development of the method of mean values in mathematical physics, and in particular in the kinetic theory of gases.

Important investigations by physicists on the foundations of mechanics are at hand; I refer to the writings of Mach..., Hertz..., Boltzmann..., and Volkman... It is therefore very desirable that the discussion of the foundations of mechanics be taken up by mathematicians also. Thus Boltzmann's work on the principles of mechanics suggests the problem of developing mathematically the limiting processes, those merely indicated, which lead from the atomistic view to the laws of continua. Conversely one might try to derive the laws of motion of rigid bodies by a limiting process from a system of axioms depending upon the idea of continuously varying conditions on a material filling all space continuously, these conditions being defined by parameters. For the question as to the equivalence of different systems of axioms is always of great theoretical interest.

If geometry is to serve as a model for the treatment of physical axioms, we shall try first by a small number of axioms to include as large a class as possible of physical phenomena, and then by adjoining new axioms to arrive gradually at the more special theories. At the same time Lie's principle of subdivision can perhaps be derived from the profound theory of infinite transformation groups. The mathematician will have also to take account not only of those theories coming near to reality, but also, as in geometry, of all logically possible theories. We must be always alert to obtain a complete survey of all conclusions derivable from the system of axioms assumed.

Further, the mathematician has the duty to test exactly in each instance whether the new axioms are compatible with the previous ones. The physicist, as his theories develop, often finds himself forced by the results of his experiments to make new hypotheses, while he depends, with respect to the compatibility of the new hypotheses with the old axioms, solely upon these experiments or upon a certain physical intuition, a practice which in the rigorously logical building up of a theory is not admissible. The desired proof of the compatibility of all assumptions seems to me also of importance, because the effort to obtain such a proof always forces us most effectively to an exact formulation of the axioms.

Suppes predicates

The trick to axiomatize a theory like classical mechanics, Schrödinger's or Dirac's quantum mechanics, or general relativity is to cradle them into an adequately rigorous formulation and then to isolate the domain of interest within ZF set theory. For, say, general relativity is (if we use a technical language) the theory of four dimensional real differentiable manifolds endowed with a Minkowskian metric.

In (nearly) everyday language: it's the theory of curved spaces of dimension four to which we add an object (the metric) that allows us to measure distances in space, and intervals of time, and which is such that we can talk about motion (velocities, accelerations) of the bodies that are on that curved space. Now with the help of a Suppes predicate we can build an axiomatic theory for these objects.

Suppes predicates arose out of a remark by Patrick Suppes in 1957:

To axiomatize a theory is to formulate a set-theoretic predicate.

In other words: to axiomatize a theory is to place it within set theory.

Newton da Costa and the late Chilean logician Rolando Chuaqui interpreted in a 1988 paper Suppes' remark with the help of ideas from Bourbaki, and came out with a simple recipe. A Suppes predicate for some theory is the conjunction of two sets of axioms (so that it becomes a single formal sentence):

- The first set of axioms describes the objects we deal with in the theory (vector spaces, topological spaces, fields, etc) and build them from scratch within Zermelo–Fraenkel set theory.

- The second set of axioms describe how these objects interlock in the theory. This set includes the motion equations, or dynamic rules, for the theory.

One has to prove that the axioms are consistent, but in general (as we deal with relative consistency) it is enough to exhibit an example or particular situation from the theory we are dealing with.

A Suppes predicate for general relativity

If we follow the Suppes guidelines in the Costa-Chuaqui formulation, we get for general relativity:

- We isolate the objects we wish to talk about: the four-dimensional curved space, the Minkowski metric (the object that gives information about distances and time intervals), velocities and the like.

- We tell the way those objects fit together and add some extra restrictions, if needed.

- We add some dynamics — motion equations — to our recipe. In the case of Einstein's general relativity these are the Einstein gravitational equations.

Done! We have our axiomatic framework for general relativity. It sits within Zermelo–Fraenkel set theory, and we can apply forcing techniques to construct set-theoretic models for it.

This kind of approach for the axiomatics of general relativity didn't come out of the blue. There was a long development that began with French mathematician Charles Ehresmann (1905–1979) who defined in 1950 the "Ehresmann connection," which can be interpreted as the gravitational field when applied to general relativity. These ideas were expanded by Polish relativist Andrzej Trautmann in the late 1960s and then formulated in a way that encompasses general relativity and unified gauge field theories by Y. M. Cho in 1975. So, there only remained to formulate everything in the cadre of axiomatic set theory and to explicitly write down the axioms we needed.

General relativity and forcing models

Once we have an axiom system for general relativity which fits it within set theory we can apply forcing techniques to it. What do we get then? There are a few quite interesting results:

- If we consider arbitrary spacetimes, an adequate forcing extension will add infinitely many new spacetimes to the already existing ones. And these "new," generic spacetimes will be different from any of the spacetimes in the model we extended by our forcing construction.

- However if we restrict ourselves to the usual Big Bang models *there will be no new Big Bang like spacetimes added in a forcing extension.*

Let us try to explain what is going on here.

A curved space is described by its coordinate charts, which are domains usually taken to be like balls that cover the whole space. The way these domains fit together ensures the particular geometry of the curved space or manifold. Now in general there are infinitely many such coordinate domains, and we can code each such manifold by a real number, say, between 0 and 1. Conversely given any real number in that interval there will be a corresponding manifold.

Now start from a model for set theory with only constructive sets. Then every real number between 0 and 1 is constructive in Gödel's sense. We then extend the original model by forcing to one with some adequate properties (we wish to have an axiom called Martin's Axiom[46] satisfied in the extended model, which is something that can always be done). Then something very unexpected happens:

> In the extended model the typical manifold will be a generic manifold. The constructive manifolds will be in a zero-probability set.

Cohen-generic spacetimes will have a 100% probability. The typical spacetimes in the forcing extension are the generic ones.

How about Big Bang universes? These are basically three dimensional bubbles that expand from the initial explosion. One such bubble is described by a finite set of coordinate domains — and forcing doesn't affect finite sets. So they remain unchanged through forcing extensions.

But — does that mean that Big Bang universes are safe from undecidability? No, just from the usual kind of forcing-generated undecidability. For we can state the following result about Big Bang universes:

> We can explicitly write down a metric tensor for a universe so that it is undecidable whether it is a Big Bang universe or a universe without a global time coordinate.

---

[46]Martin's Axiom is named after Donald Martin who formulated it in 1970 together with Robert Solovay. For our purposes it suffices to say that it implies that sets of reals with cardinalities smaller than the continuum have zero probability in the usual measure. Martin's Axiom is independent of the set-theoretic axioms.

> Moreover, for any consistent axiomatization of general relativity in Zermelo–Fraenkel set theory — indeed for a still wider, reasonable class of consistent axiomatizations of general relativity, for an explicitly built g we can write the formalized versions of the two sentences:
>
> - g is the metric tensor for a Big Bang universe.
> - g is the metric tensor of a universe without global time.
>
> Both sentences are undecidable in our theory.

Let us explain the technical concepts: a metric tensor allows us to measure spatial distances and time intervals on spacetime; it also characterizes the gravitational field. A Big Bang universe has a global time coordinate, so that it makes sense for physicists to say something like "the Big Bang happened 13.7 billion years ago." However there are universes without a global time coordinate, such as the so-called Gödel universe (more about it later).

The results above can be obtained by techniques which are different from Cohen's forcing. Wait and see.

More on forcing-dependent universes

The interpretation of those new, forcing-dependent, universes is a difficult one. A generic real isn't clearly poised in an interval; it's a kind of hazy, fuzzy object. It is as undetermined as possible, but for the fact that it is a real number and lies between 0 and 1. Similarly a generic universe is also hazy and undetermined, but for the fact that it is a universe and can be described by a covering by coordinate maps — a cover which is coded by a generic, fuzzy-like real.[47]

Yet, as we've mentioned, given an adequate forcing extension where Martin's Axiom is valid, the typical universe will be a generic universe. That means: if we collect all possible universes in a bag and pick up one at random, it will be generic with 100% probability.

But why are these new, generic universes, so counterintuitive in their looks? The whole thing is similar to the usual independence proofs of Euclid's Parallel Postulate: we redefine "point," "line," and so on, so that we may violate Euclid's Postulate. Here, for Boolean-valued models, we deal with functions that stand for sets, and given that interpretation we obtain the required properties for our generic spacetimes in a Boolean-valued set-theoretic model.

If proving that they are different from the usual, constructive universes we started from means that they will be physically different from them, then we have objects whose physical properties may reveal some surprises to us. But are they "physically meaningful"?

That's a question we wish to leave open.

---

[47] We are using "fuzzy" here without any reference to Lofti Zadeh's Fuzzy Set Theory, despite the fact that there are connections between fuzzy sets in Zadeh's sense and Boolean-valued models as used in forcing constructions.

Valentine Bargmann steps in

A closing scene. It is February 1980, and Valentine Bargmann, Einstein's last assistant and coworker, and a professor at Princeton Univeristy, readies himself to begin the first lecture in a course on general relativity at the University of Rochester, New York. The audience is quite distinguished and includes topologist Arthur Stone and his wife Dorothy Maharam Stone, a former student of von Neumann's; particle physicist Susumu Okubo, of the Okubo mass formula, mathematical physicist Gérard Emch, and many others. The first row of chairs in the auditorium is ocuppied by a few graduate students and post-docs who are there to write down in detail Bargmann's presentation (for handheld cameras and YouTube were far in the future).

Bargmann begins his lecture: "Gentlemen, I'm going to tell you not what I think about general relativity. I'm going to tell you what Einstein thought about it." And he started it.

One might wonder what Einstein would have thought of his *magnum opus* today, more than half a century after his death, from inflation to multiuniverse theories to supersymmetry and string theory. We have just offered a very small hint of the many outlandish venues that stem from his major intellectual achievement.

4. The Halting Function and its Avatars

N OW THE QUESTION IS: CAN WE DECIDE CHAOS? Is there some reasonable mathematical recipe so that, given a system of equations that describe a dynamical system, we can determine whether it will exhibit some kind of chaotic behavior?

The question is of great theoretical and practical interest. On the practical side, think of a bridge that oscillates with regularity under strong winds. Will it eventually begin to oscillate in an erratic way, and finally break down, in the midst of a storm? So, the decision problem for chaos is not just of theoretical interest; it has practical importance too.

The solution of this problem led to the construction of a function that settles the halting problem. It can be explicitly written, but it cannot be computed. We tell about it now.

Chaos is undecidable

We begin with a quote from a paper by Steve Smale:

> *Richardson and Costa-Doria made a study of [a given real-defined and real-valued function]. [. . .] We cannot decide the question, does this function have a zero in* R^n? *If one can't decide the existence of a zero, one can hardly expect to decide the existence of chaos. In fact Costa-Doria prove just that, chaos is undecidable.*[48]

It all began in 1963 when Edward Lorenz (1917–2008), a meteorologist, published a 12-page paper on "Deterministic non-periodic flow." Lorenz started out of the hydrodynamical equations which are used in meteorology modeling and simplified them in a drastic way, down to a 3-equation, 3-variable system with two innocent-looking quadratic nonlinearities.

He then tried to solve them by numerical methods, and noticed that a set of solutions behaved (or better, seemed to behave) as an impredictable well-known map called the Tent Map. But it was just an empirical remark, and it took some thirsty years to prove that the Lorenz system does in fact exhibit a kind of chaotic behavior.

This has been the curse of chaos theory since its inception. On one side we have several examples of systems that seem to exhibit some kind of chaotic

---

[48]S. Smale, "Mathematical problems for the next century," in V. I. Arnold et al., *Mathematics: Frontiers and Perspectives*, International Mathematical Union (2000).

behavior, such as the Hénon system (from celestial mechanics) or the Rössler system (from chemistry). On the other side we had several systems which were constructed with one goal in mind, that is, that they should exhibit chaotic behavior; these examples looked more or less artificial, and far from the other examples which arose from actual situations in physics. So, the main issue turned out to marry both sides of the question, namely, given a set of equations that seemed to exhibit some kind of chaotic behavior, to prove that their behavior was in fact chaotic and was due to some structure (such as the existence of a strange attractor) that was known to be associated to chaotic behavior.

Then one might ask the following question, as formulated by Morris Hirsch in 1983:

> *A major challenge to mathematicians is to determine which dynamical systems are chaotic or not. Ideally one should be able to tell from the form of the differential equations. The Lorenz system illustrates how difficult this can be.*

The answer is: there is no general algorithmic procedure to sort chaotic system from those that do not exhibit a chaotic behavior. This result is general, and independent of our characterization for chaos (there are several such particular characterizations). We may sometimes prove the existence of chaotic behavior in some systems, in a case-by-case basis, but there is no general algorithmic procedure to do the thing.

Alea iacta est

da Costa and Doria felt that chaotic systems should be somehow undecidable, or exhibit undecidable properties because in most cases their trajectories can be mapped on infinite sequences of 0s and 1s through a procedure known as a Poincaré map. They thought, if there is randomness, they thought, then there must be undecidability.

But their first efforts at the proof of the undecidability of chaos proved unfruitful for a couple of years. Then at the suggestion of da Costa, Doria took a sabbatical leave and moved to Stanford just after the 1989 Loma Prieta quake that affested the San Francisco Bay Area.

The question about chaos, however, remained always at the back of their minds as it had been the object of several discussions between da Costa and Doria. In May 1990 Suppes asked Doria to give a talk on his work, and at the end of it Suppes casually remarked, "since you are interested in undecidability in physics you should take a look at Daniel Richardson's paper on undecidable expressions in calculus, as he deals with sines and exponentials, and that might bear on quantum mechanics."

A quick walk to the math library and Richardson's paper was located. Followed an international phone call to da Costa who was in São Paulo, Brazil, and Doria could tell Suppes the next time they met: ergodicity is undecidable. Straight out of Richardson's results.

So, chaos is in fact undecidable. But there was more in stock.

Undecidability and incompleteness of chaos theory

Undecidability of chaos means that there is no algorithm to separate chaos from nonchaos; incompleteness is a more delicate affair. First of all one has to deal with an axiomatization of the theory, and prove it to be consistent (or reasonably argue for its consistency). Then one must look for some formalized sentence in the theory that looks like a candidate for independence — and prove it to be independent.

Then there was a stroke of luck: Richardson's results can be extended to an explicit expression for the halting function for Turing machines, that is, an expression for the function that solves all instances of the halting problem.

How is that possible? The halting function cannot be written within arithmetic, but can be easily written in weak extensions of it, as da Costa and Doria later noticed. How is that possible? As we will soon see, the expressions for the halting function are closely associated to Chaitin's Ω mumber, which also codes the instances when an universal Turing machine halts.

Their original construction followed Richardson, which inspired himself in an earlier suggestion by Richard Feynmann. One starts from the universal Diophantine polynomial. Feynmann had used sines and cosines to extend in a natural way Diophantine polynomials to the real numbers, and Richardson improved on Feynmann's construction. As the universal Diophantine polynomial has lets of undecidable properties, so has the Feynmann–Richardson extension.

Here Costa-Doria step in, as they noticed that one could extend the Feynmann–Richardson developments to an expression for the halting function and to a general undecidability *and* incompleteness theorem. Notice that everything was done while Doria was in Palo Alto CA and da Costa was in São Paulo, Brazil; Doria's phone bills got a sudden increase (despite the protests of his wife Margo), and Demi Getschko who was the manager of the São Paulo node of the then incipient Brazilian internet acted as a go-between: he would receive LATEX source files from Doria, typeset them and deliver them to Newton da Costa, who would then correct what was needed and fax the corrections to Pat Suppes' office at Stanford.[49]

The incompleteness theorem they proved had Post's version of Gödel's first incompleteness theorem as its blueprint. Doria finally prepared a draft for the paper on chaos and gave it to Suppes. A couple of days later Suppes called on Doria and said, "undecidability is ok, but you must be careful with incompleteness, which is a much more delicate affair, sometimes demands an argument with fast-growing functions and the like" — Suppes meant the Paris–Harrington incompleteness theorem, which had been proved ten years before. Suppes' remark was sobering, and da Costa and Doria spent two weeks in the revision of their results, until they were satisfied of their correctness. In fact what they have done is simply to code Gödel's theorem in the language of calculus and deduce incompleteness out of that coding.

---

[49]Demi Getschko is now (in 2011) one of the top-ranking executives of the Brazilian internet.

The halting function

Let's take a peek at the halting function. Actually there are infinitely many expressions for it — as there are infinitely many expressions for any function. The original construction of da Costa-Doria led to the mathematical object exhibited below.

We will need some notation. Let σ be the sign function, $\sigma(\pm x) = \pm 1$ and $\sigma(0) = 0$. The halting function $\theta(n,q)$ is explicitly given by:

$$\theta(n,q) = \sigma(G_{n,q}), \tag{1}$$

$$G_{n,q} = \int_{-\infty}^{+\infty} C_{n,q}(x)e^{-x^2}dx, \tag{2}$$

Let's explain the notation. First, consider θ:

- $\theta(n,q) = 1$ if and only if Turing machine of code (program) n stops over input q.

- $\theta(n,q) = 0$ if and only if Turing machine of code n enters an infinite loop over input q and never stops.

Finally $C_{n,q}$ is obtained from the Feynmann–Richardson transform of polynomial $p_{n,q}$, which is the two-parameter universal Diophantine polynomial. $C_{n,q}(x)$ is different from zero over the reals if and only if machine of program n stops over q. Otherwise it is zero.

But there are simpler expressions for the halting function. Here is one of those, which do not require the Richardson transforms.

Let $p(n,\mathbf{x})$ be a universal polynomial with parameter n; here $\mathbf{x}$ abbreviates $x_1, \ldots, x_p$. Then either $p^2(n,\mathbf{x}) \geq 1$, for all $\mathbf{x} \in \omega^p$, or there are $\mathbf{x}$ in ω^p such that $p^2(n,\mathbf{x}) = 0$ sometimes.

As $\sigma(x)$ when restricted to the integers is a very simple programmable function (it is what is called a primitive recursive function), we may define a function $\psi(n,\mathbf{x}) = 1 - \sigma p^2(n,\mathbf{x})$ such that:

- Either for all $\mathbf{x}$, $\psi(n,\mathbf{x}) = 0$;

- Or there are $\mathbf{x}$ so that $\psi(n,\mathbf{x}) = 1$ sometimes.

(σ of course is the sign function we've just used.) Thus the halting function can be represented as:

$$\theta(n) = \sigma\left[\sum_{\mathbf{x}} \frac{\psi(n,\mathbf{x})}{\tau^q(\mathbf{x})!}\right], \tag{3}$$

where the sum is taken over all $\mathbf{x}$, and $\tau^q(\mathbf{x})$ is an enumeration of the values of the n-variable set $\mathbf{x}$. This formulation doesn't require Richardson's trick, and looks more like Chaitin's Ω, which again is a code for the halting function.

A few technicalities

We will give some more details on the halting function as obtained out of the transformation concocted by Feynmann and modified by Richardson. In case you are not interested in that kind of detail, you can skip this section. We repeat the halting function:

$$\theta(n,q) = \sigma(G_{n,q}), \tag{4}$$

$$G_{n,q} = \int_{-\infty}^{+\infty} C_{n,q}(x)e^{-x^2}dx,$$

We now say what is within the integral sign:

$$C_{m,q}(x) = |F_{m,q}(x) - 1| - (F_{m,q}(x) - 1). \tag{5}$$

$$F_{n,q}(x) = \kappa_P p_{n,q}. \tag{6}$$

Here $p_{n,q}$ is the two-parameter universal Diophantine polynomial

$$p(\langle n, q \rangle, x_1, x_2, \ldots, x_r) \tag{7}$$

and κ_P is Richardson's transform. We are not going to explicitly describe Richardson's transform here, but we should mention a related result.

There is a function G so that:

1. *For every real number x, $G(m, x) > 1$ if and only if the halting function $\theta(m) = 0$.*

2. *There are real numbers x so that $G(m, x) \leq 0$ if and only if the halting function $\theta(m) = 1$.*

To paraphrase it: if the machine doesn't stop, then G is always over 1. If it does stop, then G dips under 0.

Why is the last result an interesting one? Because it may help us in building an analog computer that will settle instances of the halting problem that lie beyond the reach of today's comouters. Analog computers are notoriously unreliable, as they essentially draw lines with finite precision where one should have infinite precision, which is of course impossible.

Now the preceding result shows the following: if the Turing machine stops, then the associated analog-traced curve will dip into a possibly large band and go down below zero. If the Turing machine never stops, the associated line will hover beyond the band. So the "alert sign" is the drop of the line through a *band*, which can actually be made as large as one wishes, and not something that requires an infinite precision.

But this kind of analog computer has never been built. Will it work? We'll later come back to these questions.

Rice's theorem and beyond

Doria handed Suppes the draft of the chaos paper, and Suppes immediately pointed out that something very general lurked behind their arguments. That general result turned out to be a version of Rice's theorem for the language of analysis:

> Let P be some property so that one proves for some x that x has P, while for some y one is able to prove that x doesn't have property P, any such P.
>
> Then P is undecidable, that is, there is no general algorithm to decide, for an arbitrary z, whether z satisfies P or not.

Proof is easy once we have the halting function; it is enough to write,

$$z = \theta x + (1 - \theta)y, \tag{8}$$

and voilà! z is undecidable.[50]

But there is more in store: a theorem that shows that behind innocent-looking questions one might find some very nasty difficulties. We will first state it dressed up in its full technical regalia, and then explain the meaning of what's going on:

> Given some adequate axiomatic theory S which contains arithmetic, for any property P there is a x so that the formalized version of the sentence "x is in P" lies as high as one wishes in the arithmetical hierarchy.
>
> There is even some x that cannot be made arithmetically expressible within theory S.

et's now explain it. The arithmetic hierarchy is a hierarchy of difficuty steps in arithmetic. Things that are of degree **0** can de proved, even if the proof is sometimes very difficult, long and contrived. Thigs of degree **0'** can only be proved if we know how to solve the halting problem; things in degree **0''** will only be proved if we settle stuff at the **0'** level, and so on.

And there are objects that lie beyond arithmetic, that is, that cannot be made equivalent in our theory S to anything in the arithmetic hierarchy. Truly nasty objects.

Back to chaos

The result on chaos is now easy:

> There is no general algorithm to distinguish between chaotic and non-chaotic systems, for any reasonable definition of chaos.
>
> If S is a consistent axiomatization for chaos theory with enough arithmetic in it, then there is a formal sentence "x is a chaotic system" which can neither be proved nor disproved in S.

And that extends to deterministic phenomena in classical mechanics too.

---

[50]Actually as exhibited $z = z(n, q)$ is an undecidable family of such objects.

Classical mechanics is undecidable and incomplete

How about fully deterministic nonchaotic systems; do they also exhibit undecidability and Gödel incompleteness? Yes. First of all:

> *Given a consistent axiomatization of mechanics in some theory S, there is a z so that the sentence "z describes the motion of a free particle" can neither be proved nor disproved from the axioms of S.*

Construct z as follows: $z = \theta x + (1 - \theta)y$, where x is a free particle and y a harmonic oscillator (a pendulum). Ideally, this would settle the question that the authors originally conjectured, as an observation of the system described by z would show whether it oscillates or goes in a straight line. That fact had been noticed in the 1960s by Bruno Scarpellini, but in his case he dealt with — again ideal — electric circuits. Can we build them in the real world? That's an open question so far.

Costa-Doria show that there is no general integration procedure to solve the Hamilton–Jacobi equations, that is to say, there is no general algorithm to solve a problem in classical mechanics. Even if we have the solution half way, that is, even if we have a set of first integrals for a problem in mechanics, we do not have an algorithm to completely integrate it. That is to say, undecidability and incompleteness are everywhere in physics — incompleteness, of course, in the case of axiomatized systems.

How do we take these results? Are they discouraging? We wouldn't say so. Next to Chaitin's discussion of the essential randomness that lurks in the very inner workings of mathematics, they just seem to stress that the activity of mathematicians is essentially a creative one. There are no algorithmic procedures at the heart of mathematical creativity; but bo algorithms, since algorithms have only a restricted domain of application.

Mathematics is imagination.

Arnold's problems

If we were dealing with arbitrary dynamical systems, the preceding techniques would be enough to settle Arnold's decision problem for the nature of equilibrium in one such system: in the general case there is of course no algorithm to separate stable from unstable equilibria. But Arnold wanted something more specific: he restricted the consideration to polynomial systems over the rational numbers:

> *In my problem the coefficients of the polynomials of known degree and of a known number of variables are written on the tape of the standard Turing machine in the standard order and in the standard representation.*

> *The problem is whether there exists an algorithm (an additional text for the machine independent of the values of the coefficients) such that it solves the stability problem for the stationary point at the origin (i.e., always stops giving the answer "stable" or "unstable").*

I hope, this algorithm exists if the degree is one. It also exists when the dimension is one. My conjecture has always been that there is no algorithm for some sufficiently high degree and dimension, perhaps for dimension 3 and degree 3 or even 2. I am less certain about what happens in dimension 2.

Of course the nonexistence of a general algorithm for a fixed dimension working for arbitrary degree or for a fixed degree working for an arbitrary dimension, or working for all polynomials with arbitrary degree and dimension would also be interesting.

Integers were introduced in the formulation only to avoid the difficulty of explaining the way the data are written on the machine's tape. The more realistic formulation of the problem would require the definition of an analytic algorithm working with real numbers and functions (defined as symbols). The algorithm should permit arithmetical operations, modulus, differentiation, integration, solution of nondifferential equations (also for implicit functions in situations where conditions for the implicit function theorems are violated), exponentiation, logarithms, evaluation of 'computable' functions for 'computable' arguments.

The conjecture is that with all those tools one is still unable:

1. *To solve the general stability problem starting from the right hand side functions as symbols with which one may perform the preceding operations.*
2. *To solve the above problem for polynomial vectorfields with real or complex coefficients.*
3. *To solve them with integer coefficients.*

However as far as I know there are no words in logic to describe the above problem and I have thus preferred to stop at the level of algorithms in the usual sense rather than to try to explain to logicians the meaning of the impossibility of the solution of differential equations of a given type by quadratures (e.g., in the Liouville case in classical mechanics or in the theory of second order ordinary differential equations). The main difficulty here is that the solvability or unsolvability should be defined in a way that makes evident the invariance of this property under admissible changes of variables defined by functions that one can construct from the right hand side of the equations in a given coordinate system. In other terms we should explicitly describe the structure of the manifold where the vectorfield is given, with respect to which the equation is nonintegrable.

In the usual approach this structure is a linear space structure, and I think it is too restrictive.

In any case I would like to know whether you think you have proved my conjectures on polynomial vectorfields with integer coefficients:

- *For some pair (degree, dimension);*
- *For some dimension;*
- *For some degree,*

the polynomials being given on the tape of the machine in the standard form. If one of those undecidability conjectures is proved, it would be interesting to know for which pair (degree, dimension), or value of the dimension or value of the degree is the undecidability proven.[51]

Da Costa and Doria couldn't apply here the trick with the halting function, as it cannot be expressed as a polynomial. However it can be expressed as a polynomial dynamical system over the rationals, yet to be integrated, and that's how the undecidability of equilibrium states was proved. Incompleteness then follows, for adequate axiomatics.

But there is more in store. The system concocted by da Costa and Doria is such that it is undecidable whether it undergoes a Hopf bifurcation. So it is undecidable whether it has a limit cycle or not.

Let's elaborate on that. A Hopf bifurcation is a mathematical phenomenon found by Austrian mathematician Eberhard Hopf (1902–1983). The idea is: a system with a stable fixed point undergoes a transformation and becomes an oscillating system (with a stable limit circle). How is that possible? Picture yourself a marble that rolls along the walls of some vessel with a curved bottom. The marble spirals until it settles down at the bottom. Now suppose that suddenly the bottom of the vase is raised, forming a kind of hill. The marble will then spiral and instead of the final resting point in the bottom, it will go in circles around the hill at the bottom of the vase, and if the motion is frictionless, it will go forever in circular motion.

This phenomenon is found everywhere in the real world. When one has a fibrillating heart, the mild shock the heart doctor gives the patient makes the heart restart its regular beat, that is, makes it undergo a Hopf-like transition from a (nearly) stable condition to a normal beating state (the stable cycles). One can also find it in economics.

So, as the Hopf bifurcation is undecidable in the Costa-Doria system, then it is undecidable whether it has a limit cycle or not. Therefore one settles Arnolds 1974 problems.

The halting function, the halting problem and the Ω number

In 2007 Cris Calude published a volume to honor Greg Chaitin on his 60th anniversary: *Randomness and Complexity: From Leibniz to Chaitin.*[52] In the introduction he mentions a result by Robert Solovay on the Ω number:

There is an Ω number none of whose digits can be computed in ZF.

Costa and Doria saw that remark, and proved two related theorems:

There is a computable real number none of whose digits can be computed by ZF.

---

[51]Communicated by e-mail in a message from Arnold to Costa-Doria.
[52]World Scientific, 2007.

For any "nice" theory S with enough arithmetic, there is an Ω number none of whose digits can be computed by S.

We now tell the trick, as in those TV shows where a magician does something mysterious on the stage and then discloses how to perform it.

- For the first result. Let σ be a one-sided infinite computable binary sequence; then $s = .\sigma$ is a binary irrational. Let σ^* be the complementary sequence, where 0 is changed to 1 and vice versa, and let $s^* = .\sigma^*$ be the corresponding complementary binary irrational. Then put:

$$x = s\theta + (1 - \theta)s^*.$$

  x is computable, because it is either s or s^*; however θ is such that ZF cannot decide whether it is 0 or 1. Therefore while we know that x is computable, as it equals one of the computable irrationals s or s^*, we cannot decide the actual value of each digit in x as we cannot decide θ. So, we are done.

- For the second result there are actually two proofs, and the thing is a bit more convoluted. A "nice" theory is one whose theorems can be enumerated by some computer program. If our nice theory S contains enough arithmetic,[53] then each non-halting instance of the halting problem for a given fixed universal Turing machine is given by a Π_1 sentence, that is, by a formal sentence that translates as "for all x, x has property P," where deciding whether x has P can be computationally done.

  Sorry for the technicality. Then we may suppose that there is a listing of all those instances of the halting problem, $\xi_1, \xi_2, \xi_3, \ldots$, even if we cannot actually compute that listing. (Why is that so? These are non-halting instances of the halting problem which are beyond the reach of S; if we could compute them, we would be able to concoct a proof for them in S, since we can internalize the computation as a proof in S.) We then use that listing to define our version of Ω.

  Of course as we cannot prove each ξ_i in S, we cannot compute (in S) any digit of our new Ω.

  Again we are done. There is still another proof that requires the axiom of choice, but we will skip it.

Economics and the social sciences

Newton da Costa has always had students from the economics department at the University of São Paulo. One of them was economist Fernando Garcia, who proved in his PhD thesis that if one uses a paraconsistent logic — a logic

---

[53] We also suppose that S has a model where the arithmetic part is standard, that is, given by the usual integers.

that admits controlled contradictions — then the argument in the celebrated Arrow's Impossibility Theorem doesn't go through.

Arrow's impossibility result is also known as the No-Dictatorship Theorem: under reasonable assumptions, there is no way a voting system would give results that follow the dictates of a single voter (the dictator). The central point in the proof is that the argument reaches a contradiction. But Garcia showed that for an adequate paraconsistent logic we can frame that contradiction in such a way that it becomes an "acceptable" contradiction in our logical background, and so Arrow's argument won't go through.

In 1991 Doria wrote a short article to *Jornal do Brasil*, a newspaper in Rio, where he briefly mentions his joint results with da Costa and speculates on their application to economics. In a few days Newton da Costa was approached by former Brazilian minister of finance Delfim Netto who asked him if he could take a look at some work by one of his students (Delfim Netto was a professor of economics in São Paulo). The student was Marcelo Tsuji, a Brazilian of Japanese stock that had just received his B. Sc. in economics and was considering the alternatives for a PhD in some related field.

Costa, Doria and Tsuji met twice in May 1991. In the second meeting Tsuji sat with da Costa and Doria in the small coffee lounge at the department of philosophy where da Costa taught, and sketched to them how to apply the halting function to obtain an undecidability result for Nash games and for competitive markets. That took ten, fifteen minutes, for Tsuji had already filled in the main details of the argument.

Let's elaborate on what Tsuji did.

Nash games

Game theory was conceived first by Ernst Zermelo,[54] of set theoretic fame, and then had a decisive contribution by Émile Borel, a French mathematician (1871–1956), who conjectured in the 1920s what is now known as the "minmax theorem." Then John von Neumann step in and gave a proof of that theorem, which appears in full in his great work (with Oskar Morgenstern) *Theory of Games and Economic Behavior*, published in 1944.

However there was a catch with von Neumann's result, and that's where Nash enters the picture.

Everybody has an intuitive idea of what a game is:

- A game has n players.

- A game has rules.

- A game is a succession of moves by each player according to its rules.

- A game has some kind of payoff which the players receive at the end of the gane.

---

[54]FAD must thank his friend Vela Velupillai for his guidance on the origins of game theory.

John von Neumann following in Borel's steps proposed a concept of solution for a game, the so-called minmax solution. But there was a difficulty: von Neumann proved that every 2-person game has a minmax solution, but generalization to n-player games proved very difficult. Then John Nash suggests in his 1950, 27 page (!!!) PhD thesis passed at Princeton a new kind of solution for a n-person game, the Nash equilibrium:

> *The Nash equilibrium is such that no strategy used by any player can diminish the payoff received by any one of the participants in the game; a change of strategy either leave things as they stand or improves them.*

Nash proves that every n-person, competitive game always admits a Nash equilibrium. (A game is competitive if no coalitions or collusions among players are allowed.) It was this result which was honored by the 1994 Nobel Prize in economics. It is central to the theory of competitive markets, as it is a key step in the argument of Kenneth Arrow and Gerard Debreu's famous 1952 paper on "Existence of equilibrium for a competitive economy."

The theory of competitive markets had been formulated by mathematical economist Léon Walras in the late 19th century. Walras conjectured that competitive markets always had equilibrium prices, that is, prices where supply and demand are equal. Yet he couldn't prove it, and his conjecture remained open for more than half a century until Arrow and Debreu settled the score in their joint 1952 paper, duly honored by the 1972 Nobel Prize in economics. Nash, whose 1950 theorem provided a decisive point in the Arrow-Debreu proof, had to wait for 22 more years for recognition.

Competitive markets and fixed-point theorems

A fixed-point theorem is of the form:

> *Given a set A and a map F from A onto itself, F has a fixed point if there is some x_0 in A so that $F(x_0) = x_0$.*

Fixed-point theorems appear everywhere. Get yourself a cup of a hot latte, open it, move it with a spoon. You'll notice that the center of the liquid remains stable while the rest moves. Here F is the movement imparted by the spoon, and x_0 is the center of the surface of your latte.

Market stability is equivalent to the existence of a fixed point. Let $o(p), d(p)$ be the offer and demand functions, where p is the variable that describes prices. Then we have equilibrium if the offer equals the demand:

$$d(p) = d(p). \tag{9}$$

Now make a few transformations:

$$o(p) - d(p) = 0. \tag{10}$$

$$o(p) - d(p) + p = p. \tag{11}$$

Given $F(p) = o(p) - d(p) + x$, we conclude that $F(p) = p$. That is, market equilibrium is equivalent to the existence of a fixed point for F. So, in order to prove that there are equilibrium prices you just have to browse a catalog of fixed-point theorems, look for the best fit for F, and apply the chosen flavor of fixed-point to it. That works for sure.

However the beauty of the Arrow-Debreu result is that it interpretes equilibrium prices as something that arise out of a game, and out of its solution as a Nash equilibrium. Nash equilibria are conceptually very easy to understand:

> *A Nash equilibrium is such that any change in strtegy by the other players of the game will either leave things as they are or will improve on your gains.*

So, it's not just a set of equilibrium prices that you get; it's something whose intuitive nature you can easily figure out.

The Tsuji result

The Tsuji result was originally motivated by earlier, similar, results by Alain Lewis (we'll just comment on them). The trick is like the way one proves the undecidability of chaos. Let Γ' and Γ'' be two competitive games with the same number of players and with payoffs given by u', u''. Then game Γ with payoff function:

$$u = \theta u' + (1 - \theta)u'' \tag{12}$$

is undecidable and leads to an incompleteness theorem:

> *For a consistent axiomatization of game theory, the sentence "game Γ has payoff function $u = u'$" is undecidable.*

We are going to elaborate later on that. This result didn't come out of the blue; it was preceded by extensive work of Alan Lewis and Vela Velupillai.

Alain Lewis, Vela Velupillai

1991, May. It's time of the late late night show on tv. Doria would be fast asleep at home. But the phone would ring and wake him up (and his wife too) and a deep, almost *basso profondo* voice would greet him on the other side: Alain Lewis. And a hour-long discussion of undecidability and incompleteness would follow until Doria declared himself tired, and Lewis would concede and finish the conversation.

Sometimes huge packets with hundreds of handwritten pages would come in the mail for Doria. Drafts of papers by Lewis, copies of his letters sent by Lewis to top-ranking economists such as Kenneth Arrow and to Gerard Debreu. Lots of math stuff, usually both rigorous and deep. Technical results mixed up with conceptually brilliant stuff on the foundations of economics.

Lewis is sometimes nicknamed "the other Nash." He was born in Washington D.C. in 1947, got his PhD in economics at Harvard and had an erratic career with troubles similar to the ones that plagued John Nash. His main work has to do with the foundations of mathematical economics and undecidability issues — what one usually calls "metamathematical phenomena." Briefly, a sample of Lewis' work:

- Lewis' chief result on the undecidability of game theory had been announced since 1986, but only appeared in print in 1992, even if it had been widely circulated beforehand. The result is, in technical language: recursively presented games have a non-recursive arithmetic degree.

  We explain it: even if we can describe a game by some recursive function, its equilibria may be noncomputable. This is a remarkable result, which inspired the work by Tsuji, da Costa and Doria. (We'll just comment on the differences between both results.)

- A second construction, more impressive yet, is the rather unexpected discussion where Lewis shows that the Busy Beaver function naturally appears in the theory of choice functions. It is again a remarkable result, that may tie in with the fact that the counterexample function to the $P = NP$ hypothesis grows in its peaks at least as fast as the Busy Beaver function.

Other papers by Lewis are concerned with the rather surprising independence of formulations of an old conjecture in economics, Edgeworth's Conjecture, with respect to Zermelo–Fraenkel (ZF) set theory or with the effective content of scientific theories in general — with an eye on economics. (Edgeworth's Conjecture was formulated by Francis Y. Edgeworth in the 19th century. Briefly and waving hands it says that as the number of agents in an economy increases towards infinity, indeterminacy is reduced. More precisely: when the number of agents tends to infinity, even if we allow for cooperation among agents, the resulting situation tends to a Walras-like equilibrium.

Surprisingly Lewis shows that for some reasonable formalizations of the conjecture, the formal statement that corresponds to it can neither be proved nor disproved within axiomatic set theory.)

Vela Velupillai's work was introduced to Newton da Costa and Chico Doria again by the former Brazilian economy minister Delfim Netto, who sent them one of Vela's preprints in 2003, and asked for their comments (Vela's paper dealt with constructive aspects of the mathematics behind economics). da Costa urged Doria to contact Vela, and so Doria did in the same year. It happened that Vela had read an essay written by Costa-Doria in 1995 whose title was "Variations on an original theme," which is a tongue-in-cheek reference to Sir Edward Elgar's celebrated *Enigma* Variations (Vela Velupillai has a very refined musical taste).

Let's now take a peek at Vela's work. It is aptly summarized in his 2000 book *Computable Economics*. Briefly:

- The book *Computable Economics* sketches the main research lines in the field. A full chapter is voted to the approach that identifies the rational behavior of an economic agent with the behavior of a Turing machine.

  Then follows a central result: if we identify rational agents with Turing machines, we get that preference orderings are undecidable. Such a simple but remarkable result is proved as a direct consequence of the unsolvability of the halting problem for Turing machines. Later chapters in the same book describe the concept of Turing computability and its effect on game theory and introduce some intriguing new explorations in computable economics. We stress that this result by Vela is the first example of a naturally occurring undecidable situation in economics.

- The paper "The unreasonable *in*effectiveness of mathematics in economics" which was published in 2005 is a detailed discussion of the role of mathematics in economic modeling. Velupillai sums it up with a remark by Sir John Hicks (1904–1989), that the concepts of accounting lie at the center of economics. If we are then to deal with integer-valued entities and arithmetic relations among those entities, we are in the domain of the theory of Diophantine equations, which is rigged with undecidable questions.

- In "Algorithmic foundations of computable general equilibrium theory" (2006) Velupillai proves another key result: the usual theory of the computable general equilibrium is neither computable nor constructive in the strict mathematical sense. His discussion is based on the theory of computer programs.

The Velupillai results and examples are framed as *undecidability* results, that is, as results that show the nonexistence of algorithms to settle some specific questions. However they can lead to *incompleteness* results, that is, results that exhibit sentences ξ in the language of some adequate consistent axiomatic theory such that neither sentence ξ nor its negation can be proved within the given axiomatic framework. For instance, given a "nice" theory S such as the one referred to here, with enough tools to formulate the concepts we deal with, the result stated in p. 50 of *Computable Economics*, namely:

> There is no algorithm to decide whether a given class of decision rules is a steady state of some adaptive process.

leads to the following:

> If S is consistent and has a model with standard arithmetic for its arithmetic portion, for some α, a class of decision rules, the sentence "α is a steady state of an adaptive process" is such that neither it nor its negation can be proved within S's axiomatic framework.

There are several other authors who have contributed to the field, e.g. Kislaya Prasad, who also dealt with the incompleteness of Nash games, or Stefano Zambelli, who modelled several procedures in economics through Turing machines. The whole field of Turing machine theory is covered in their essays and research papers, and we should especially note that the Busy Beaver function again comes up in several flavors. There was also a recent review of the field by J. Barkley Rosser Jr and Roger Koppl (see the references).

So, there is already a well-established line of research in the field of mathematical economics that investigates the effects of undecidability and incompleteness in it, and its possible practical consequences, which we now briefly mention.

The meaning of the Lewis–Tsuji result

What is the meaning of the Lewis undecidability theorem for finite games? And of Tsuji's analogous result?

- The Lewis result says that for competitive games where players and outcomes are described by some computer program, the Nash equilibrium is undecidable. This immediately translates to a competitive market where participants are also described by a computer program; its equilibrium prices are undecidable.

  (Of course, if we explicitly give payoff matrices for the game, and if the description of the players is simple enough, then we can compute the Nash equilibria.)

- The Tsuji result supposes that the game, and the associated market, are described in the language of classical analysis, as in most economic models. Then we get undecidability, in the general case.

Simple cases are not expected to lead to undecidability. But — can we say that economic situations are simple?

Can we predict the future of historical events?

In 1951 Isaac Asimov (1920–1992) published a novel on a mathematician who was able to predict the course of history. That book is *Foundation*, and it became the first volume of an extended saga on the evolution of a galactic empire long in our future. The mathematician's name was Hari Seldon, and he describes his technique for predicting the future as "my little algebra of humanity." According to Asimov, it would mix up higher-dimensional geometries and calculus.

The current models for the evolution of biological and social systems are based on dynamical systems — systems of equations that describe the time-evolution of some process — and were originally inspired by the way we describe chemical reactions in the field of chemistry kinetics. Chemical models were used by Alfred Lotka (1880–1949) who was a physical chemist by training, to describe ecological system. His best known model are the Lotka–Volterra equations (named after Lotka and Vito Volterra (1860–1940), an Italian mathematician); they describe simple predator-prey systems.

The Lotka–Volterra equations were independently conceived by Alfred Lotka and by Vito Volterra, and have several interesting features:

- They exhibit *nonlinear cycles*, that is, they are nonlinear equations with cycle solutions.

- They apply to several examples of predator-prey systems. For instance, we can use them to model the predator-prey behavior of deer and their predators (say, big cats) in national parks.

  We can also use them to model epidemic cycles, where the predators are the bugs, and we are their prey. Or we can use it to model a politically incorrect situation in economics: trade unions prey on the gains of corporations, and may lead to economic cycles.

There are many interesting features in the Lotka–Volterra system, but as everything that has been presented so far, we can easily concoct undecidable Lotka–Volterra systems with the help of the halting function θ. And we cannot eliminate the possibility that, for very complicated systems, some "natural" undecidable behavior will creep up. And of course complicated systems are what we get in economic and in the social sciences.

The Lotka–Volterra system was first applied in economics by Richard Goodwin, so instructs us Vela Velupillai.

Forcing, the halting function and Ω

The last chapter concluded with a few paragraphs on Cohen's forcing; here we elaborate on several applications of the halting function to the proof of undecidability and incompleteness theorems.

Now we may ask: is there a relation between Cohen's forcing and our proofs of undecidability and incompleteness out of expressions for the halting function? We may say so, and we will now sketch a few ideas about that possible connection between both techniques.

Recall the picture we gave of a Cohen generic real between 0 and 1: we require it to have the property of being a real number between 0 and 1, and all properties that are either equivalent to that or that follow from it in set theory, and none other. So, we know about Cohen-generic reals that they lie between 0 and 1, and nothing more. They are sort of fuzzily placed within the prescribed interval — and we know nothing more about them.

We will now sketch a construction that will lead to an object that is as fuzzy as a Cohen real, but which uses the halting function θ:

- Recall that we can write down a Turing machine program that enumerates all theorems of ZF set theory.

- Start that machine, and select all theorems that describe properties P of real numbers between 0 and 1 such that there are different real numbers r, s so that P holds for r, and doesn't hold for s.

- We can build in several ways a sequence of independent θ functions, $\theta, \theta', \theta'', \ldots, \theta^{[n]}, \ldots$.

The actual expression for our desired fuzzy object leads to a rather involved expression, which can be seen in a paper by Costa–Doria, "Variations on an original theme," published in 1996.

Nevertheless that object is as fuzzily placed between 0 and 1 as any Cohen-generic real. It is a real number that lies between 0 and 1 — and nothing more can be said about it.

Ω and θ

We can now come full circle and link together our main themes, Ω, θ — and the Busy Beaver function. This comes out of a simple and elegant suggestion by Chaitin. Pick up one of those versions (may we say avatars, pace James Cameron?) of the halting probability Ω so that ZF cannot compute a single digit of it.

Represent as follows its decimal expansion:

$$\Omega = .\theta_1 \theta_2 \theta_3 \theta_4 \theta_5 \ldots \cdot \qquad (13)$$

Then we get a new fuzzy, θ-generic, now Ω-generic real number z^Ω: just by using the same expression we had for the θ-generic object described above.

The ourobouros has bitten its tail, and we have come full circle. Ω and θ are entities that represent the same phenomenon: the halting problem. And so is the Busy Beaver function: if we solve the halting problem, we can compute the BB function. If we know the BB function, we can solve the halting problem.

The sacred serpent has bitten its tail...

5. Entropy, P vs. NP

W E NOW BEGIN TO EXPLORE UNCHARTED TERRITORY. In this chapter we advance two conjectures, one of them about Shannon's theorems in information theory, and the other one on the nature of the P vs. NP question.

Entropy, random sequences, the Shannon theorems

The Shannon coding theorems were proved in the late 1940s by Claude E. Shannon (1916–2001), an electric engineer from the Bell Labs. The concepts we require to understand the meaning of the Shannon theorems are:

- *Entropy or information quantity.* A quantitative measure for the intuitive concept of information. Suppose that you are in front of a set of alternatives, and that you know the probability of choosing each alternative. The measure we are describing gives you the amount of information you gain when you make a choice and reduce the uncertainty you had before making that particular choice.

  The information quantity is modeled after the concept of entropy from statistical mechanics and thermodynamics.

- *Channel capacity.* Think of the information channel as a pipeline. If it is narrow, the information amount through it will be restricted, as with pipelines that carry water. The channel's capacity is the measure of how narrow a channel is (or how broad it is).

The Shannon coding theorem says that for channels with a small amount of noise we can find a coding procedure for our message pool so that the messages will be transmitted as close to the channel's capacity as one wishes, and also with as little mistakes as desired.

The Shannon Coding Theorem: a closer look at it

We can split Shannon's results into two situations: first, what happens to noiseless channels. Then what happens when we deal with noisy channels. For noiseless channels, the situation is quite simple. Recall that a communication channel connects a *source* to a *receiver*. The communication process itself is a map that preserves probabilities. Thus the information amount (or entropy) at the receiving end cannot be smaller than the information amount (or entropy) of the source. In symbols:

$$H_{\text{source}} \leq H_{\text{receiver}}. \tag{1}$$

We are done. However for noisy (weak noise!) channels, things are much more involved. Let's now look at the proof of Shannon's theorem for noisy channels. We have long noticed that a simple, straightforward presentation of the argument behind Shannon's great result is lacking; this is the main motivation of the present section. We have followed here Shannon's original ideas, since they are pretty clear; we then comment on a few features of it.

This is the argument's backbone:

1. *The Shannon–McMillan–Breiman theorem.* This is the most difficult step in the argument. It is a very technical and elaborate result, and it is assumed without proof in Shannon's original 1948 paper. (It is motivated by some much simpler particular cases.)

   We state it here and argue for its plausibility.

2. *The equipartition argument.* This is the central idea. It is derived from the Shannon–McMillan–Breiman result and shows that, given some conditions, we can substitute entropy for probabilities, and have neat estimates for the size of sets of messages.

   That's a crucial point: entropies become here proxies for probabilities. We can use entropies in the place of probabilities, with good approximation.

3. *The kernel of the argument.* The whole thing comes from the following question: which is the probability that a message in the receptor has no corresponding message in the source? (Is it therefore a mistaken message? Or did it exist already in the source?)

   We will see that this probability tends to 0, even in the presence of (weak) noise, for adequate coding.

The whole argument only requires elementary algebra, with a little touch of calculus, but for the Shannon–McMillan–Breiman step. But it is still a technical argument. We give it here for the sake of completeness, as it isn't found outside technical texts. If you wish, you can skip it and go straight to page 95.

The Shannon–McMillan–Breiman theorem

Recall that the "average event" $E(X)$ defined by the probability-weighted average:

$$E(X) = \sum_i p_i X_i \tag{2}$$

is the *expected value* of variables $X_1, X_2, \ldots, X_n$ with the corresponding probabilities $p_1, p_2, \ldots, p_n$.

Now define the *entropy per sign* of a set X_n of messages of length n:

$$h(X_n) = (1/n)H(X_n). \tag{3}$$

We consider the limit $n \to \infty$ (one must prove that the limit exists; however we assume it here without much ado). Notice that:

$$h(X_n) = E\left[-(1/n)\log p(s_n)\right]. \tag{4}$$

(Here $p(s_n)$ denotes the probability of sequence $s_n \in X_n$.) Define $h(X) = \lim_{n\to\infty} h(X_n)$. For very large n then,

$$h(X) \approx E\left[-(1/n)\log p(s_n)\right]. \tag{5}$$

The Shannon–McMillan–Breiman theorem asserts that for *ergodic, stationary* processes, one actually has (nearly always),

$$h(X) \approx -(1/n)\log p(s_n). \tag{6}$$

This is a result that greatly simplifies our computations: for nearly all sequences, and large n, $p(s_n)$ becomes independent of s_n. Moreover, we can drop the expectation value operator in the calculation of the actual value of the entropy.

The equipartition property

It can be stated as:

> Let X be a set of messages of very large length n and entropy per letter $h(X)$. Then X splits into two subsets:
>
> 1. A very large subset $X_L \subset X$ of total probability ≈ 1 (or 100%). X_L has approximately $2^{nh(X)}$ messages, each of probability approximately equal to $2^{-nh(X)}$.
> 2. A very small subset $X_S = X - X_L$ of total probability ≈ 0.

Here is the argument for it. Recall that the log is base-2 logarithm. From the Shannon–McMillan–Breiman theorem,

$$h(x) \approx -(1/n)\log p(s_n).$$

This is the same as:

$$p(s_n) \approx 2^{-nh(X)}.$$

This happens "for almost all" s_n. Therefore their number is approximately $2^{nh(X)}$.

This is a very important property! It means that we can use information instead of probabilities, almost always.

The Shannon Coding Theorem: final steps

It can be stated as:

1. *Suppose that we have a communication system with source X, receptor Y, connected by a channel of capacity C.*

2. *Let $h(X)$ and $h(Y)$ be entropies per letter of the source and of the receptor.*

3. *We suppose that there is a weak noise in the channel.*

Suppose moreover that the transmission rate R of messages through the channel is very close to C, that is, $C - R = \eta$, η a very small positive number.

Then there is a coding for the source so that the error in the transmission of the messages is as small as one wishes.

The argument: we have the entropy of the source $h(X)$, the entropy of the receptor $h(Y)$; we know that the noise $h_Y(X)$ is weak. Therefore, given the equipartition theorem, we can estimate how many messages are in the source and in the receptor.

The goal is to compute the size (cardinality) of the set of the wrong messages in the receptor, that is, the set of messages in the receptor that *do not* originate in the source.

Then:

- There are $2^{nh(X)}$ messages in the large probability group in the source.

- There are $2^{nh(Y)}$ messages in the large probability group in the receptor.

- Transmission speed is $R = C - \eta$, η small. Messages transmitted are $\approx 2^{nR}$, according to the equipartition property.

- The probability that a message of length n is in the source's high probability set is:

$$2^{nR}/2^{nh(X)} = 2^{n(R-h(X))}. \tag{7}$$

This is: the number of (nearly all) transmitted messages divided by (nearly all) messages in the source.

Now: *which is the probability that no message in the source becomes a received message?*

- The probability that a message *doesn't originate* in the source is, roughly:

$$1 - 2^{n(R-h(X))}. \tag{8}$$

- There are $2^{nh_Y(X)}$ messages in the receptor.

- Notice that the probability that *no message in the receptor* originates in the source is the probability that *one* message doesn't originate in the source times itself as many times as there are messages in the receptor.

- Then *the probability that no message in the source becomes a received message* is

$$P = (1 - 2^{n(R-h(X))})^{2^{nh_Y(X)}}. \tag{9}$$

Now: channel capacity (plus our approximation hypotheses) lead to the following:

$$R + \eta = C \approx h(X) - h_Y(X), \tag{10}$$

$$R - h(X) \approx -h_Y(X) - \eta. \tag{11}$$

The probability above computed becomes:

$$P = (1 - 2^{-nh_Y(X)-n\eta})^{2^{nh_Y(X)}} \tag{12}$$

Now:

- *Noise is weak.* Thus $2^{nh_Y(X)} \approx 1 + nh_Y(X) \log 2$, since $2^x = e^{x \log 2}$, and $e^{ax} \approx 1 + ax$.

- η is a very small positive constant. Therefore:

$$2^{-nh_Y(X)-n\eta)} \approx 1 - n \log 2(h_Y(X)) - n\eta \log 2. \tag{13}$$

If we substitute those expansions above we get:

$$P \approx 1 - (1 - n(h_Y(X) \log 2 - \eta \log 2 + h_Y(X) \log 2))$$
$$+ \text{quadratic terms in } h_Y \text{ or in } \eta h_Y. \tag{14}$$

Simplification gives:

$$P \approx 1 - (1 - n\eta \log 2) \approx 1 - 2^{n\eta}. \tag{15}$$

n is large but fixed. For $\eta \to 0$, P tends to 0.

Randomness in a game of heads in tails

Back to less technical stuff. Think of a game of heads and tails, in all its possible outcomes. Those outcomes can be represented by infinite sequences of 0s and 1s, e.g. 011010001 ... and it goes on forever. For an arbitrary game of heads and tails, all such sequences must be considered, and there are $2^{\aleph_0}$ of them, that is, they have the cardinality of the continuum. We can calculate its entropy per digit, and is value $h = 1$.

The meaning of that value is quite intuitive: given our infinite game of heads and tails, when we move from play n to play $n + 1$ we gain exactly one bit of information.

However — whenever one deals with infinities, get ready for conterintuitive results. Proceed as follows:

- Exclude from our game all sequences that end in infinite strings of zeros and ones, that is, those that are 0000000... or 1111111... but for a finite initial segment.

- Exclude from our game all computable sequences, that is, those that can be generated by some computer program (they include the preceding case).

What happens then to the entropy h? Nothing! It is still equal to 1. This comes from the following theorem:

> Given an infinite game of heads and tails as we have just described, we can exclude from it an arbitrary zero-measure set without affecting the entropy per digit h.

(A zero-measure set is a set of zero% probability.)

We now ask the following question: *where does the entropy reside?* Which is the set of, say, truly random sequences that generate the non-zero value for the entropy? In order to answer that question we'll have to resort to a few tricks — and we'll see that the answer we get is quite surprising. We proceed as follows:

- Axiomatize everything within ZF set theory. Consider Gödel's constructive universe and obtain within it our infinite game of heads and tails. Then every sequence in our game is constructive in the sense of Gödel, even if random.

- Now enlarge the constructive universe by an adequate forcing procedure, and ask that Martin's Axiom be valid in the extension.

  We have that the original set of constructive sequences can still be recognized within our extended model, but *has become a zero-measure set.* The value of the entropy results from the new generic sequences that have been just added our game when we made the forcing extension; the original random set contributes nothing to the entropy in the enlarged model.

So, *where does randomness reside?* Why is it that the set that originated the value of the entropy in our game now gives no contribution to it?

There is another consequence here: we can isolate the set $\hat{x}$ that contains the constructive sequences in the exttended model. In the original, constructive model, the entropy per sign $h(\hat{x}) = 1$. In the extended model we have that $h(\hat{x}) = 0$. The information content of $\hat{x}$ is trivialized.

Now recall Shannon's coding theorem for communication channels:

> Given a communication channel with weak noise, of capacity C, there is a coding for the messages being transmitted so that they can be sent through the channel with rate as close as one wishes to C, and similarly with as few errors as desired.

If the entropy of the message set exceeds C, then errors may be arbitrarily large. The idea is:

> Given a communication channel of capacity C with a set of messages of entropy $h > C$, is there a model extension that allows us to change that inequality into $0 < h < C$ so that Shannon's theorem applies?

We now that there is one such change that trivializes the information content, which we have just mentioned. Is the nontrivial situation feasible?

That question is open, but it might be of great practical interest — if it is positively answered, and moreover if it can be implemented in the real world in some reasonable, nontrivial, way. It would perhaps provide a way of using narrowband communication channels to transmit messages from a pool of large entropy.

P vs. NP

The gentle lady who acts as secretary of a college department asks one of the professors: you know math, don't you? Then can you help me in one of my chores? I have to distribute timetables for classes, teachers and students in such a way that there is a minimum overlap among them as prescribed by our regulations.

And it is always so hard to get a solution... The only way I know to do it is to try all possible combinations until I find one that fits. And that takes a long long time. Is there a simple way of doing it?

The professor smiles. For the lady is asking him: is $P = NP$?

Before we explain the notation, let's consider another example. The most famous problem in the NP class is the *traveling salesman problem* : given N cities connected by a web of roads, which is the shortest route among them? Which route is such that it has length at least equal to, or larger than, some fixed, constant value l ?

In the first case we have an example of a NP-hard problem; in the second case, of a NP-complete problem. But we can treat both kinds simultaneously.

Another example is the family of *allocation problems* as above. We have sites geographically dispersed over some region, and scarce resources to be distributed among them. We want to do it efficiently (for some previously established criteria of efficiency) and again can either impose an optimizing condition (which will make it into a NP hard problem) or fix some efficiency level (the NP complete case). Or we wish to solve the teachers-and-classes distribution problem, as the lady in the college department asked the professor.

A brief history of problems in the NP class

The problem goes a long way back; it is nearly two centuries old. We can trace the traveling salesman problem back to Sir William Rowan Hamilton in the

19th century, who formulated it as a question about what we now know as "Hamiltonian circuits," which is an alternative way of looking at the traveling salesman. (Hamilton was a very bright fellow: he stated the Minimum Action Principle for classical mechanics, formulated mechanics in what we now know as the Hamiltonian picture, and discovered the quaternions — which Wolfgang Pauli used to add the concept of spin of an electron to Schrödinger's wave equation.)

The first recent formulation of the traveling salesman problem (TSP) is usually credited to mathematician Karl Menger (1902–1985) in 1932. Menger's formulation of the TSP is quite straightforward:

> *We call the Messenger Problem [. . .] the task of finding, for a finite number of points whose pairwise distances are known, the shortest path connecting the points. The problem is naturally solvable by making a finite number of trials. No rules are known that would reduce the number of trials below the number of permutations of the given point.*

The name "traveling salesman problem" appears nearly two decades later, in a 1949 report prepared by Julia Robinson (yes, she's the one who contributed to the solution of Hilbert's Tenth problem) for the RAND Corporation.

Then follows Gödel's much-quoted letter to von Neumann in March 1956 where the problem is again formulated, now in the context of a Boolean satis-fiability problem.

Cook and Karp (see Machtey and Young's book for references and details) characterize and list NP-complete problems in the early 1970s; they are seen to pop up everywhere in both concrete and abstract situations.

One may now state the $P = NP$ question in two versions, the informal version and the more formal version:

> Informal version. *Is there a fast way of solving the traveling salesman problem in all cases?*
>
> More formal version. *Is there a polynomial algorithm that settles all instances of some NP-complete problem?*

Polynomial Turing machines and related fauna

Input some binary sequence, say, a k-bit sequence, to a Turing machine. It is said to be a polynomial Turing machine if the time it takes to process that input and to output some stuff, is bounded by $k^n + n$, for some fixed positive n. (The operation time of a Turing machine is the number of cycles it takes to process some input until it outputs something, if ever.)

More technically: recall that Turing machines are supposed to input sequences of 0s and 1s such as, say, 001010. They are called *bit strings*. A *polynomial Turing machine* (or *poly machine*) is a Turing machine whose operation time is bounded by some polynomial on the length of the input string. (The length of a string is the number of 0s and 1s in it; given a bit string x, its length is noted $|x|$, and the bound may be noted $|x|^n + n$.)

Now let's explain the name of the problem. P stands for "polynomial," and NP for "nondeterministic polynomial." Think of the traveling salesman problem. Suppose that given N cities, there are some $N!$ possible routes to check. If there is a Turing machine that branches out and performs $N!$ computations in paralell, at the same time, the obvious algorithm for TSP is to test each one of the possible $N!$ routes with that massively parallel machine. That can be made as fast as one tests for a single route, and one can easily see that this is time-polynomial (suffices to add the length of the connecting roads between adjacent cities).

The NP class of problems and the $P = NP$ conjecture

We insist: NP stands for "nondeterministic polynomial"; we'll comment again on its meaning of that at the end of this paragraph. Problems in the NP class are described by the slogan:

> *If you know the solution for a problem in NP then you can test (check) it very fast. However if you don't know the solution then it is very hard (it takes a long long time) to get one solution in the general case.*

More precisely: they are easy to check because they can be checked by a poly machine (a time-polynomial machine). They are hard to find because in the general case nobody knows an algorithm for it which is polynomial, i.e., can be implemented by a poly machine. Then follows the $P = NP$ question (this equation means, NP problems are solvable by poly machines):

> *Is there a poly machine that settles all instances of NP problems?*

Recall that the so-called nondeterministic poly (that is, $NP-$) machines can settle any NP problem in polynomial time, as described above, whereas the use of NP to denote that class of problems.

If there is no such an algorithm, then $P = NP$ is false. We write $P < NP$ for the negation of $P = NP$.

Shared thoughts

One of the interesting facts of scientific research is its generous side. When it became known that Newton da Costa and Chico Doria were interested in the $P = NP$? question, several generous researchers approached them and offered to share their wildest thoughts about the problem. Most of them did so under the condition of anonymity, and they usually communicated hints like, "see if you can prove fact X," or "have you noticed that property Y holds?" There was also much help on the side of proof-checking; Marcel Guillaume, from Clermont–Ferrand, offered in 2000 to go in detail through the Costa–Doria arguments and see if they held water.

Their first published paper on the matter appeared in 2003 in a journal that dealt with more technical, not foundational stuff, and which was then edited by John Casti, *Applied Mathematics and Computation*. It was purposely written in a very dry style, with proofs as formal as possible and nearly without comments, and contained a conditional result of the form:

> *If condition A happens, then* ZF *cannot prove* $P < NP$.

Condition *A* essentially translates: if such-and-such Turing machines behave as they should in the real world of integers etc, then ZF cannot prove $P < NP$.

Their result received an immediate and scathing review at a noted journal; in fact it was published before the paper appeared in print. (Of course that kind of review can always be taken as a kind of *compliment à outrance*.) The conditional theorem was called a "gap" in the argument — but no other difficulties were found with it. So, the whole thing was correct.

The crucial intuition?

The current view in computer science is that $P < NP$ will eventually be shown to hold. Newton da Costa and Chico Doria share that belief — with one caveat: which formal system is strong enough to prove $P < NP$? Would Peano Arithmetic be strong enough to prove it? Most researchers believe that it would, since the problem can be entirely formulated within arithmetic (more about that in a moment). Would we need the much stronger tools which are available in the ZF toolbox?

The two crucial facts about P vs. NP that arose in the informal discussion between Costa–Doria and other researchers are:

- If $P = NP$ and $P < NP$ are independent of ZF, and if ZF has a model where arithmetic is standard (we must assume it: that means, arithmetic is as in the real world of bank accounts, financial statements, and real-life computers) then $P < NP$ is true.

  That means: Peano Arithmetic plus an infinitary rule prove $P < NP$.

- The nonrecursive "counterexample function" to $P = NP$ grows in its peaks faster than any total recursive (total computable) function. That is, it behaves kind of like the Busy Beaver function.

(List all poly machines: the first instance, if any, of a problem in the NP class, which some poly machine fails to settle, gives a value for the counterexample function; see below.)

Why are these folklore-like facts important? First, not every arithmetic sentence can be proved solely within arithmetic. In fact there are arithmetic sentences that lie beyond the reach of ZF:

> *We can write down the programs for an infinite family* Φ *of Turing machines, so that the formal, fully arithmetic, version of the sentence "Φ*

is a family of poly Turing machines" can neither be proved nor disproved within ZF.

In fact we may have to add lots of weird mathematical animals such as inacessible cardinals galore to ensure that that sentence may be proved.

Also the way ZF "sees" the Busy Beaver function is highly distorted, in the sense that the Busy Beaver function tops all naïvely total recursive functions, while ZF can only "see" those functions which are ZF–*provably* total recursive. So, the full Busy Beaver function sort of lies outside the reach of ZF, and remains so even if we strengthen it with large cardinal axioms. (Which are the looks of the Busy Beaver from within ZF? It is F_{ZF}, an intuitively total recursive function which cannot be shown to be so within ZF. That is to say, very far from the original Busy Beaver function.)

So, both folklore-like facts about the P vs. NP question are relevant to our pursuit. The following has also been pointed out: suppose that to our system S, admitted to be consistent, we add an infinite collection of axioms:

$$N > 0$$

$$N > 1$$

$$N > 2$$

$$\ldots$$

for a natural number N. We get a consistent augmented system, and N is said to be a *nonstandard, finite, large natural number.* It is bigger than any natural number that we can explicitly write down. And, again, we must stress, it is consistent to add all those new axioms to an already consistent system like our S. Then write down the following polynomial:

$$p(x) = 1 + \frac{x}{1!} + \frac{x^2}{2!} + \frac{x^3}{3!} + \cdots + \frac{x^{N-1}}{(N-1)!} + \frac{x^N}{N!}. \tag{16}$$

(The dots ... mark the moment, so to say, where one begins to have numbers which depend on N.) Before the dots we only have standard natural numbers — in fact, we remain within a model for S with only standard numbers. Beyond the dots we have a model with nonstandard numbers. Waving hands,

- When we restrict ourselves to standard natural numbers, $p(x)$ becomes an exponential:

$$e^x = 1 + \frac{x}{1!} + \frac{x^2}{2!} + \frac{x^3}{3!} + \cdots \tag{17}$$

- When we include what lies beyond the dots, we see that $p(x)$ is a nonstandard polynomial of degree N.

If we take $p(x)$ to be the operation bound for some Turing machine, it will be seen to look polynomial when nonstandard integers are allowed, but exponential when only standard integers come into play.

The main result: the counterexample function to P = NP grows too fast

We wil now have to be somewhat more technical. For details and a complete proof see the reference list at the end of the book. Suppose that the negation of $P = NP$, that is, $P < NP$ holds. (That means: it is true in the standard model for arithmetic.) Then for each poly machine to which we input a set of strings that, each one, codes some problem in the NP class, there will necessarily be a moment when the machine gives the wrong answer to the input string.

That is to say, the counterexample function is total. Now, some technicalities.

More precisely: we can code problems in the NP class by binary strings ordered in such a way that they are themselves represented by the natural numbers $0, 1, 2, 3, \ldots$. We then input those strings in that order to some poly machine P_n (where n codes the program of the machine as a Gödel number). If it first fails for input k we put that the value of the counterexample function $f(n) = k$. (If n doesn't denote a poly machine then put $f(n) = 0$.)

Always suppose that $P < NP$ holds. Then:

> *The counterexample function f grows in its peaks faster than any total function which can be computed by some Turing machine.*

Functions that can be computed by some Turing machine are called *partial recursive functions*. A function is total whenever it is defined for every input. So our result states that f grows in its peaks faster that any total Turing-machine programmable function. We can also say that f overtakes any total Turing-machine programmable function infinitely many times.

The idea of the proof is simple. Pick up an arbitrary, very steep fast growing total recursive function f (a total function which is computable by a Turing machine). Then there is a fact:

> *We can explicitly construct a poly machine that gives correct answers to all instances of a problem in the NP class up to a fixed chosen k, and then fails forever.*

Then for an adequate and faster-growing g, construct a family of such machines that fail after outputs $g(1), g(2), g(3), \ldots$, so that the values of the counterexample function to $P = NP$ will be $g(1) + 1, g(2) + 1, g(3) + 1, \ldots$. For an adequate choice of g, these values overtake those of any f as above.

Trivially:

> *If the counterexample function f to P = NP is total, that is, if P < NP holds, then it proves that any intuitively total recursive function is in fact total.*

For we can code arbitrary naïvely total recursive functions into f by the procedure described above, and so if f is total, that is, if $P < NP$ holds, then every segment of f is total over its restricted domain.

The main conjecture: if P < NP is true then it cannot be proved by reasonable axiomatic systems

We must say what we understand by "reasonable axiomatic system." It is a system like PA (Peano Arithmetic, axiomatized arithmetic with finite induction) or ZFC (Zermelo–Fraenkel set theory, strengthened with the axiom of choice). These systems share the following characteristics:

- If consistent, we may suppose that they have a model with standard (usual) arithmetic.

- They have a set of theorems which may be generated by a Turing machine (we say that the theorems are *recursively enumerable*).

(There are other, more technical, conditions here, which we skip for the moment; see the references.) These reasonable theories, so to say, have the following already mentioned property:

> Given a reasonable theory S there is a total recursive function F_S which cannot be proved total by the axioms of S.

We can construct F_S in such a way that it dominates all provably recursive total functions in S. So F_S grows faster than any S-provably total recursive function. Now the conjecture:

> No consistent reasonable theory S proves that the counterexample function f is total.

Could we then argue that it would then prove F_S total recursive, as a segment of f? If so, no such theory S would then prove P < NP (modulo our hand waving; for details and loopholes see the references).

Construction of F_S is easy as we now review it: we enumerate with the help of a computer program all theorems of S, pick up those that say f_i is total, for some i that codes our enumeration, and diagonalize over those functions, that is, we define:

$$F(i) = f_i(i) + 1, \tag{18}$$

where by construction F is different from any f_i. There are other constructions where one immediately sees the fast-growing property of that function.

If P < NP is independent of S, then it is true of the standard integers

This is an important result which however results from a simple argument. Suppose that a clever computer programmer concocts a poly algorithm (an algorithm that can be implemented by a poly machine) for all NP problems and writes it out in full. Suppose that an integer a codes that wonderful program. Then the following is true:

> For every instance x of an NP problem, poly machine P_a outputs a correct answer to it.

That sentence when properly formalized becomes what is known as a Π_1 sentence, and it is known that when we add a true arithmetic Π_1 sentence to a theory like PA (Peano Arithmetic) very little is changed. Thus pick up PA and add that (supposedly true) sentence to it. We get a theory that proves $P = NP$. Either trivially as we have it as an axiom, or nontrivially, as PA might itself prove it. Anyway both theories share the same language and the same "provability strenght," that is, they have the same provably total recursive functions, as the addition of any true Π_1 arithmetic sentence to PA doesn't change the set of provably total recursive functions in the theory.

So, if $P = NP$ is true then either PA or a theory very much like PA proves it.

Conversely, if either PA or a theory like it (as above) does not prove $P = NP$, then it cannot be true of the standard model for arithmetic. Follows that if both $P = NP$ and $P < NP$ are independent of PA or even ZFC, or of any theory like the S we used before, then $P < NP$ holds true of the standard integers, that is, of the real world where we find computers among other things.

Another discussion

More technicalities. Costa and Doria originally published a different argument, which also appears in their more recent papers. Given a theory like S, they pick up function F_S and use as bounds for the poly machines, polynomials like $|x|^{F_S(n)} + F_S(n)$. A few computations show that S proves the equivalence:

$$[P < NP]^F \leftrightarrow ([P < NP] \wedge [\text{F is total}]). \tag{19}$$

Here $[P < NP]^F$ is our modified definition, with the subscript S dropped. Also one has that S proves:

$$[P < NP]^F \rightarrow [\text{F is total}]. \tag{20}$$

Follows that S cannot prove $[P < NP]^F$, since S cannot prove that F is total. However that doesn't settle the issue, as we must have that S doesn't prove $P < NP$, and it can only be derived given the equivalence:

$$[P < NP] \leftrightarrow [P < NP]^F \tag{21}$$

and we can show that such an equivalence is Gödel-independent of the axioms of S, supposed consistent.

But do we really have independence?

We now use a trick that first appeared in a famous 1975 paper by Baker, Gill and Solovay. Pick up an arbitrary Turing machine and plug to it a polynomial clock that counts the steps in the computation and shuts down the machine the moment it is about to exceed the bound computed by the clock. That arrangement is also a Turing machine, one that we can ensure is a poly machine, from its construction. The set of all those machines is the BGS set; it contains representatives of every conceivable poly machine (that is, it contains machines that emulate all known poly machines).

So far the arguments exhibited point towards the strong possibility that a theory like S, which subsumes PA and ZFC, does not prove $P < NP$. To prove independence we require another hypothesis:

> If S proves that $P = NP$, then the set of BGS poly machines that decide *every instance in a NP-problem is recursive in theory S.*

(Recall that a set is recursive if there is a computer program that allows us to decide whether some element x is in that set or not; the BGS set is a convenient way to represent all poly machines, as it is a recursive set — however it doesn't include all poly machines, only sort of representatives for each collection of machines that perform the same calculations; see the next section.) The argument exhibited by Costa–Doria in support of that hypothesis is simple, but again incomplete: one turns on the machine that lists the theorems of S and separates all theorems of the form "integer n codes an algorithm for a machine in the BGS set that solves all instances of a NP-problem." One then should conclude that the listing is exhaustive.

Then, given that hypothesis, we can concoct a version f' of the counterexample function which again grows faster than any total recursive function, and the conclusion follows: S cannot prove $P = NP$ too.

More precisely: if $P = NP$ is true, then it can be proved so by a reasonable theory, as we have seen. Then the counterexample function will be undefined at the values of n that code algorithms that settle, for such machine will never fail. The conjecture asserts that the set of undefined values of f, given that $P = NP$ holds, is recursive. We can therefore easily fill in the holes and obtain a fast-growing function f' that grows faster than any total recursive function etc.

And we get the independence of $P = NP$ and of $P < NP$ from strong, reasonable, axiomatic systems at the end of our discussion.

Still more conjectures on the counterexample function

$P < NP$ is given by a Π_2 arithmetic sentence, that is, a sentence of the form "for every x there is an y so that $P(x, y)$," where $P(x, y)$ is a very simple kind of relation.[55] Now given a theory S with enough arithmetic in it, S proves a Π_2 sentence ξ if and only if the associated Skolem function f_ξ is proved to be total recursive by S. For $P < NP$, this function is what we have been calling the counterexample function.

However there are infinitely many counterexample functions we may consider, an *embarras de choix*, as they say in French. Why is it so? For many adequate, reasonable theories S, we can build a recursive (computable) *scale of functions* $F_0, F_1, \ldots, F_k, \ldots$ with an infinite set of S-provably total recursive

---

[55] It is a primitive recursive predicate.

functions so that F_0 is dominated by F_1 which is then dominated by $F_2, \ldots$, and so on.

Given each function F_k, we can form a BGS-like set BGS^k, where clocks in the time-polynomial Turing machines are bounded by a polynomial:

$$|x|^{F_k(n)} + F_k(n), \tag{22}$$

where $|x|$ denotes the length of the binary input x to the machine. We can then consider the recursive set:

$$\bigcup_k BGS^k \tag{23}$$

of all such sets.

Each BGS^k contains representatives of all poly machines (time polynomial Turing machines). Now, what happens if:

- There is a function g which is total provably recursive in S and which dominates all segments f_k of counterexample functions over each BGS^k?

- There is no such an g, but there are functions g_k which dominate each particular f_k, while the sequence $g_0, g_1, \ldots$ is unbounded in S, that is, grows as the sequence $F_0, F_1, \ldots$ in S?

In the first case, S proves $P < NP$, and we are done. However the second case leads to an interesting nontrivial situation which is so far unexplored. We believe we can use here techniques similar to those that we applied in the study of the full counterexample function.[56]

---

[56]See F. A. Doria, *Chaos, Computers, Games and Time*, E-PAPERS/PEP/COPPE (2011).

6. Forays into Uncharted Landscapes

I S REALITY REAL? German philosopher Martin Heidegger (1889–1976) concludes both his 1929 talk *What is Metaphysics* and begins his 1953 treatise *Introduction to Metaphysics* with the question: *Why is there anything, instead of nothing?*

Heidegger deems that question to be the central issue of metaphysics. We may add to it our own query:

Is reality real?

where we have in mind the current opposition between "real" and "virtual," an opposition which didn't exist when Heidegger published his treatise, and people still referred to the big mainframe computers of the 1950s as "electronic brains."

Why do we ask those questions?

In 1999 Roland Emmerich produced, and Josef Rusnak directed a sci-fi movie, *The Thirteenth Floor*, which explicitly raises the issue of how real is reality. The movie's plot stems from a 1964 novel by Daniel Galouye, which seems to have been one of the first science-fiction novels to talk about a fully simulated virtual world. (There is a precedent in *The City and the Stars* (1954), by Arthur C. Clarke, but virtual reality isn't the central motive of its plot as it happens in *The Thirteenth Floor*.)

In the Emmerich–Ruznak movie the story begins in Los Angeles, 1999, where a computer programmer developed a full virtual reconstruction of Los Angeles in 1937. The movie ends in Los Angeles in the 2030s where one learns that the 1999 reality was also a simulation. And we leave the movie theater with the question: isn't the 2030 reality again some kind of virtual reality, some computer simulation?

So, we go back to our question: *is reality real?*

We can still mention another version — should we say avatar? — of our question. It was explicitly formulated in 1936 by Kurt Riezler (1882–1955), a German politician who became a philosophy professor and intellectual mentor of a whole generation of American philosophers. Riezler published in 1936 an essay with a rather neutral title, "The metaphor in Homer and the beginnings of philosphy." However the opening lines of Riezler's essay clearly states a version of our problem:

> *The poet shows and makes comprehensible the soul as the world, and the world as our soul. Their unity is his secret.*

Riezler talks about "inner world" and "outside world," an opposition that seems to mirror the pair virtual reality/concrete reality. So, ours is an old, respectable philosophical question. What can we say about it, given our discussion in the preceding pages? There is a serendipitous coincidence here, as Riezler's essay appeared in print in the same year that gave us the Church, Kleene and Turing papers on the concept of computation.

Let's frame our main question here in a language closer to our ideas: *is the universe "pure" software?* In order do deal with it we propose the following road map:

- What is software? How far can we stretch its concept? Or, should we deal with a more restricted semantic domain for 'software'?

  We'll discuss here the problem of hypercomputation, or super-Turing computation.

- Time and space. The hardness of any computation is discussed in terms of the amount of time it takes, or the amount of space it requires.

  Can we give an unconventional view of these concepts?

Then — we try to sum it up.

Is the world built out of information?

Is everything software? Now for some even weirder stuff! Let's return to *The Thirteenth Floor* and to the ideas that we briefly referred to in the introductory section of this chapter.

Let's now turn to ontology: What is the world built out of, made out of?

Fundamental physics is currently in the doldrums. There is no pressing unexpected, new experimental data — or if there is, we can't see that it is! So we are witnessing a return to pre-Socratic philosophy with its emphasis on ontology rather than epistemology. We are witnessing a return to metaphysics. Metaphysics may be dead in contemporary philosophy, but amazingly enough it is alive and well in contemporary fundamental physics and cosmology.

There are serious problems with the traditional view that the world is a space-time continuum. Quantum field theory and general relativity contradict each other. The notion of space-time breaks down at very small distances, because extremely massive quantum fluctuations (virtual particle/antiparticle pairs) should provoke black holes and space-time should be torn apart, which doesn't actually happen.

Here are two other examples of problems with the continuum, with very small distances:

- *the infinite self-energy of a point electron in classical Maxwell electrodynamics,*

- *and in quantum field theory, renormalization, which Dirac never accepted.*

And here is an example of renormalization: the infinite bare charge of the electron which is shielded by vacuum polarization via virtual pair formation and annihilation, so that far from an electron it only seems to have finite charge. This is analogous to the behavior of water, which is a highly polarized molecule forming micro-clusters that shield charge, with many of the highly positive hydrogen-ends of H_2O near the highly negative oxygen-ends of these water molecules.

In response to these problems with the continuum, some of us feel that the traditional

Pythagorian ontology:
God is a mathematician,
the world is built out of mathematics,

should be changed to this more modern

→ Neo-Pythagorian ontology:
God is a programmer,
the world is built out of software.

In other words, all is algorithm!

There is an emerging school, a new viewpoint named **digital philosophy.** Here are some key people and key works in this new school of thought: Edward Fredkin, http://www.digitalphilosophy.org, Stephen Wolfram, *A New Kind of Science,* Konrad Zuse, *Rechnender Raum* (Calculating Space), John von Neumann, *Theory of Self-Reproducing Automata,* and Chaitin, *Meta Math!*.[57]

These may be regarded as works on metaphysics, on possible digital worlds. However there have in fact been parallel developments in the world of physics itself.

Quantum information theory builds the world out of qubits, not matter. And phenomenological quantum gravity and the theory of the entropy of black holes suggests that any physical system contains only a finite number of bits of information that grows, amazingly enough, as the surface area of the physical system, not as its volume — hence the name *holographic principle.* For more on the entropy of black holes, the Bekenstein bound, and the holographic principle, see Lee Smolin, *Three Roads to Quantum Gravity.*

One of the key ideas that has emerged from this research on possible digital worlds is to transform the **universal Turing machine**, a machine capable of running any algorithm, into the **universal constructor**, a machine capable of building anything:

---

[57]Lesser known but important works on digital philosophy: Arthur Burks, *Essays on Cellular Automata,* Edgar Codd, *Cellular Automata.*

Universal Turing Machine → Universal Constructor.

And this leads to the idea of an *information economy*: worlds in which everything is software, worlds in which everything is information and you can construct anything if you have a program to calculate it. This is like magic in the Middle Ages. You can bring something into being by invoking its true name. Nothing is hardware, everything is software![58]

A more modern version of this everything-is-information view is presented in two green-technology books by Freeman Dyson: *The Sun, the Genome and the Internet*, and *A Many-Colored Glass*. He envisions seeds to grow houses, seeds to grow airplanes, seeds to grow factories, and imagines children using genetic engineering to design and grow new kinds of flowers! All you need is water, sun and soil, plus the right seeds!

From an abstract, theoretical mathematical point of view, the key concept here is an old friend from Chapter 2:

$H(x) =$ the size in bits of the smallest program to compute x.

$H(x)$ is also $=$ to the minimum amount of algorithmic information needed to build/construct x, $=$ in Medieval language the number of yes/no decisions God had to make to create x, $=$ in biological terms, roughly the amount of DNA needed for growing x.

It requires the *self-delimiting programs* of Chapter 2 for the following intuitively necessary condition to hold:

$$H(x,y) \leq H(x) + H(y) + c. \tag{1}$$

This says that algorithmic information is sub-additive: If it takes $H(x)$ bits of information to build x and $H(y)$ bits of information to build y, then the sum of that suffices to build both x and y. Furthermore, the mutual information, the information in common, has this important property:

$$H(x) + H(y) - H(x,y) = \begin{cases} H(x) - H(x|y^*) + O(1), \\ H(y) - H(y|x^*) + O(1). \end{cases} \tag{2}$$

Here

$H(x|y) =$ the size in bits of the smallest program to compute x from y.

This triple equality tells us that the extent to which it is better to build x and y together rather than separately (the bits of subroutines that are shared, the amount of software that is shared) is also equal to the extent that knowing a minimum-size program y^* for y helps us to know x and to the extent to which knowing a minimum-size program x^* for x helps us to know y. (This triple equality is an idealization; it holds only in the limit of extremely large compute times for x and y.)

---

[58] On magic in the Middle Ages, see Umberto Eco, *The Search for the Perfect Language*, and Allison Coudert, *Leibniz and the Kabbalah*.

These results about algorithmic information/complexity H are a kind of economic meta-theory for the information economy, which is the asymptotic limit, perhaps, of our current economy in which material resources (petroleum, uranium, gold) are still important, not just technological and scientific know-how.

But as astrophysicist Fred Hoyle points out in his science fiction novel *Ossian's Ride*, the availability of unlimited amounts of energy, say from nuclear fusion reactors, would make it possible to use giant mass spectrometers to extract gold and other chemical elements directly from sea water and soil. Material resources would no longer be that important.

If we had unlimited energy, all that would matter would be know-how, information, knowing how to build things. And so we finally end up with the idea of *a printer for objects*, a more plebeian term for *a universal constructor*. There are already commercial versions of such devices. They are called 3D printers and are used for rapid prototyping and digital fabrication. They are not yet universal constructors, but the trend is clear...[59]

In Medieval terms, results about $H(x)$ are properties of the size of spells, they are about the complexity of magic incantations! The idea that everything is software is not as new as it may seem.

Hypercomputation: or where are the limits of software?

Formal treatments for the concept of computability and of computable function slowly developed in the 1930s since Gödel's great paper on the incompleteness of formalized arithmetic. Other landmarks are Alonzo Church's paper on the λ-calculus, Kleene's paper on general recursive functions, and Turing's great paper on his mathematical machines. All of them dated around 1936, a kind of *annus mirabilis*.

The Church–Turing thesis arose in that context. It cannot be proved, as it asserts the equivalence between a heuristic, informal concept, that of calculability or computability, and formal constructions shown to be equivalent, such as Turing machines, general recursive functions, and the λ-calculus:

The Church–Turing thesis

A function is intuitively computable if and only if it is computable by a Turing machine.

Or there is a general recursive computation procedure for it, or there is a λ-calculus procedure, etc. Other equivalent formalizations for the concept of computability as crystallized in the Turing machine picture are Markov algorithms, Post's procedure, and even cellular automata, whose universality as a computation device was proved by Alvy Ray Smith III in 1972 (he since moved to the movie industry where he became famous by his work on the animations of Star Trek's *The Wrath of Khan* and by his work at Pixar).

---

[59]One current project is to build a 3D printer that can print a copy of itself; see http://reprap.org

One of the main arguments in support of the Church–Turing thesis is the fact that mathematical proofs can be formalized as computer programs, that is, as Turing machines. But can all mathematical arguments be thus reduced?

No.

Consider the following example of a naïvely acceptable mathematical argument. First, pick up the Turing machine that enumerates all proofs in set theory (ZF axiomatic set theory). Then:

- Begin the enumeration of all proofs in ZF.

- You've reached proof coded n. Pick up among all preceding proofs those that say, "f_e is a computable function of program coded by e and f is total," that is, it has no bugs. (We can do that in an acceptably computable way.)

- Compute all values for those functions up to argument n. It is a finite set. Then chose its largest value. Let it be N.

- Define function $f(n) = N + 1$.

Of course f is different from any f_e that appears in the enumeration just described, as its value exceeds by 1 at least, that of any total computable function up to n as described. So, the proof of f being a total function cannot appear among the ZF proofs, if ZF is consistent.

Now: how can we prove that f is a total function?

We can argue as follows: for each n, f has a well-defined value. Then, f must be total. OK? Most people would accept this as a naïvely satisfactory argument. But let's make it explicit:

- $f(0)$ is defined.

- $f(1)$ is defined.

- $f(2)$ is defined.

- $f(3)$ is defined.

- ...

- Then, for every n, $f(n)$ is defined.

This looks naïvely correct. Yet, it is an infinitary rule, which cannot be mechanically implemented, for how are we to write down in finite time that infinite list of arguments?

This is the kind of question that Turing took up in his 1939 thesis, when he tried to expand arithmetic in such a way that it became complete.

On hypercomputation

We may formulate the *hypercomputation problem* as follows: is there a real-world device that settles questions which cannot be solved by a Turing machine? We can also understand hypercomputation as as the theory (and possible applications) of systems whose behavior includes some non-algorithmic physics. That is, we may say that we are trying to understand the behavior and computational power of physical models that cannot be simulated by computer programs.

We will not describe in detail the development of hypercomputation theories — for historical details see the 2006 special issue on hypercomputation of *Applied Mathematics and Computation*, a journal then edited by John Casti. In that issue Martin Davis and Mike Stannett address at the beginning the problem of hypercomputation: will it work? Is it an absolute impossibility, an impossible dream?

We just mention a few landmark results in what can be seen as the prehistory of hypercomputation theories.

- *The concept of computable function.* As mentioned already, is usually recognized as originating in three almost simultaneous papers by Church, Turing, and Kleene. The equivalence among the three formulations was immediately proved by Kleene and Turing, and they establish our current concept of computation through the so-called Church–Turing thesis.

- *Hypercomputation theory.* Or super-Turing computation. It is the theory of any device or devices whose calculating properties exceed those of any Turing machine, or equivalently which exceeds the computing power of partial recursive functions, formal constructs which derive from the above-mentioned papers by Kleene, Church and Turing. In short: a hypercomputer computes more functions than a Turing machine, or it computes a Turing uncomputable function, e.g. it computes the (uncomputable) halting function.

- *The first abstract hypercomputing device.* The first theoretical hypercomputer that appears in the literature is Turing's 1939 oracle machine. The now quite well-known idea is to couple a new, unspecified device — the oracle — to a Turing machine; at some steps in a calculation, the oracle is consulted, and the computation only proceeds after the oracle's answer. As it is well-known, some oracle machines are strictly more powerful than the usual Turing machines (or partial recursive functions) in the sense that they compute more functions than Turing machines. That is, they are hypercomputers.

- *Progressions of theories.* Turing's 1939 paper provides a full theory of hypercomputation through a central result that was later expanded by Feferman in what is now known as the Turing–Feferman theorem, a result whose consequences and implications are still poorly understood. We will elaborate on that later on.

- *Other hypercomputation theories.* They include Spector and Addison's $\aleph_0$ mind — a device that is regulated by a finite set of instructions, operates deterministically, in discrete steps, but is able to eventually carry out a search through an infinite sequence of steps, if required. More recently Dana Scott introduced a novel theory of computation which is strictly stronger that Turing machine theory, but it doesn't seem to be translatable into some plausible physical device.

We must also mention the Blum–Shub–Smale theory of computation, which generalizes Turing machine theory into a theory of computation over the reals. Again it doesn't seem to be able to be implemented by some physical device, and in fact it doesn't actually exhibit hypercomputational behavior.

Analog computers as ideal hypercomputers

So, hypercomputation theory started just after the appearance of the standard Church–Turing computation theory (CT). CT theory is the theoretical backbone behind the PCs of today. Now the question is: can we *build* some concrete hypercomputing machine that stands to some hypercomputation theory in a relation similar to that of our PCs to CT computation theory?

To our knowledge one of the very first remarks in that direction is Scarpellini's rather surprising observation in a 1963 paper that:

> Analog devices can decide some undecidable arithmetic sentences.

(We'll later come back to the quote in Scarpellini's paper.) In fact, specific ideal analog devices can decide all undecidable arithmetic sentences, as we will soon discuss. They will decide for their truth in the standard model for arithmetic, that is, the concrete world of computers and of everyday calculations.

However Scarpellini's statement was surprising since analog computers were seen as a kind of poorer, deficient relatives of digital computers. Yet same idea briefly considered by Scarpellini was again examined by Georg Kreisel in 1976.

A possible hypercomputer

da Costa and Doria have degrees in engineering. Engineers build things. So, our goal here is to sketch a series of steps at whose conclusion we would have an actual, working hypercomputer.

Will it work? We leave that question unanswered. But, as we insist, we are cautiously optimistic that its answer may turn out to be a "yes."

We go back to Scarpellini's 1963 paper, where we find our main idea:

> In this connection one may ask whether it is possible to construct an analog-computer which is in a position to generate functions $f(x)$ for which the predicate $\int f(x)\cos(nx)\,dx > 0$ is not decidable while the machine itself decides by direct measurement whether $\int f(x)\cos(nx)\,dx$

is greater than zero or not. Such a machine is naturally only of theoretical interest, since faultless measuring is assumed, which requires the (absolute) validity of classical electrodynamics and probably such technical possibilities as the existence of innitely thin perfectly conducting wires. All the same, the (theoretical) construction of such a machine would illustrate the possibility of non-recursive natural processes.

(Scarpellini's paper discusses among other results the decidability of such predicates.) Now, we may reformulate what we said before and say that the hypercomputation problem splits into two questions:

1. *The theoretical hypercomputation problem.* Can we conceive a hypercomputer, given ideal operating conditions?

2. *The practical hypercomputation problem.* Given a positive answer to the preceding question, can we build a concrete, working, hypercomputer?

We argue here that the answer to the first question is a definite "yes," if we accept that ideal analog machines fit into the requirements ("ideal operating conditions" — just take a look at Scarpellini's papers on the matter; they are available on the web). The second question boils down according to our viewpoint to an engineering problem; we may answer it with a "maybe," or "we have to see." Or, if we follow the classical injunction: build a prototype!

Then we will see how it performs; which engineering problems must be overcome in order to have a decently working hypercomputer. For in principle it can be built.

Prototype for a hypercomputer

We have shown in a previous chapter that we can write down an explicit expression for the halting function for Turing machines in the language of calculus. We will use here a Turing machine coupled to an analog device that settles the halting problem, according to Scarpellini's prescription and with the help of the Costa–Doria results on the halting function. So, the hypercomputer might work as follows:

- Start the coupled Turing machine + analog device.

- Input the data entered to the Turing machine to the analog device. It will tell whether the machine stops or enters an infinite loop given that data set.

- If it gives a 'stop' sign, input the data to the Turing machine and perform the computation.

- If not, abort the operation.

The hypercomputer and true arithmetic

True arithmetic is complete arithmetic, a theory where everything true is provable. Such a theory can be formalized as, say, Peano Arithmetic plus an infinitary rule, like Hilbert's ω-rule, or Shoenfield's weaker but still powerful recursive ω-rule (see the Franzen paper in the references). Its theorems, how- ever, cannot be listed by a Turing computer program, as they form a nonrecur- sive set of sentences, which no computer program will enumerate in its totality.

Yet our proposed hypercomputer can be adapted and extended to work as a kind of "theorem proving (hyper)machine" that settles all arithmetical truths.

Such a device goes beyond the solution of the halting problem; it can be extended in such a way that it decides "higher order truths" of any arithmetical sentence. More technically, every arithmetical sentence lies in a given "degree of unsolvability," and the sequence of degrees is noted $0, 0', 0'', 0''' \dots 0'$ codes a sequence of 0s and 1s that can be computed. $0'$ codes the halting problem. $0''$ codes the higher-level halting problem for a Turing machine with an oracle for the halting problem. And so on, for $0''', 0''''$, indefinitely, for all integers.

Now Costa–Doria showed that there are higher-order versions of the halt- ing function for all those degrees; they can also be explicitly written out. If we can (hyper) compute them, we can decide the truth of arithmetic sentences at their level. This means: our proposed hypercomputer makes arithmetic into a complete theory. Truth and provability coincide here.

More on the theory of hypercomputation

Actually one requires very little to settle the halting problem. There is a simple result that shows it. Suppose that $I_1, I_2, \dots, I_n$ are true non-halting instances of the halting problem, that is, say, I_1 means, "Turing machine T enters an infinite loop when fed input x_1." Then:

> Given any finite set $I_1, \dots, I_n$, of true non-halting instances of the halting problem, then there is an algorithm that settles those instances.

Any finite set of non-halting instances. (The proof is simple: arithmetic + $I_1, \dots, I_n$ proves true those instances, trivially. And the proof is a Turing algo- rithm — albeit a trivial one — for those particular instances of the problem.) Why can't we collect those particular algorithms into a single one? Basically because such a rounding up of particular cases would require an infinitary argument, something which is not allowed in our concept of Church–Turing computability.

We now go back to something that was briefly mentioned before, the con- cept of "progressions of theories." That concept goes back to Turing's thesis, which was published in 1939. He asked: given the enumeration of theories — *PA* here is Peano Arithmetic — and Consis(PA), the Gödel sentence that asserts the consistency of *PA*, etc:

- *PA* + Consis(*PA*).

- $[PA + \text{Consis}(PA)] + \text{Consis}[PA + \text{Consis}(PA)]$.

- ...

and proceeds up to as far as possible in the enumeration of those hierarchies of theories, what does he prove of extra about arithmetic sentences?

The answer: suffices to go up to ordinal $\omega + 1$, and one proves all true non-halting instances of the halting problem. (Roughly, if we pick up PA and add to it the sentences that say that all naïvely total – bugless — computer programs are in fact total, then we get a theory that proves all true arithmetical sentences along the so-called arithmetical hierarchy.)

Recent research

Several papers and some books on hypercomputation, theoretical and applied, have appeared in the last two decades. We notice the review of the Church–Turing Thesis by Coleman. Hypercomputation has also been the main topic of special issues of journals, e.g. *Minds and Machines* and the already mentioned issue of *Applied Mathematics and Computation*.

There are also internet resources on the subject, such as Mike Stannett's "Hypercomputation Research Network" (http://www.hypercomputation.net) which has been available since 2001; several of the researchers listed in Stannett's homepage are regular contributors to the question. Books on the subject are still scarce; one notes Havah Siegelmann's *Neural Networks and Analog Computation: Beyond the Turing Limit*, and *Superminds*, by Bringsfjord and Zenser, as well as Apostolos Siropoulos' *Hypercomputation*, which appeared in 2008.

Hypercomputation is about some theoretical limit given by the boundary between the computable and the hypercomputable. We will now discuss another barrier, very likely a more hallowed one; nearly sacred: the universal limit given by the speed of light in the vacuum.

Spacetimes: exotic variations on that theme

We measure the complexity of computations by the amount of space and time they require. So, in our exploratory efforts, why don't we go deep into that conceptual jungle and take a look at the ideas about space and time, from an unusual viewpoint?

We are going to present here a *continuous* viewpoint of nature — and will later make a heretical suggestion of how to make it compatible and coexistent with the discrete view we get from quantum mechanics.

Exoticisms

It all came into being in a 1956 paper by John Milnor, where he proved that the 7-dimensional sphere S^7 admitted "nonstandard," — exotic, according to Milnor — differentiable structures, all compatible with its topology, which makes it a sphere.

Let's elaborate on it. A topological manifold is a curved space, from the viewpoint of mathematics. It is defined by a set of charts — collection of local maps with coordinates, and rules to piece everything together. We require those maps to be continuous, and we get a topological manifold as a result.

If moreover we wish to make the manifold into an arena for some physical processes, we must describe velocities and accelerations on it. This requires the introduction of an extra structure in our manifold: the charts must also be differentiable. Now when we describe an ordinary, standard manifold, we give it an atlas (a collection of charts) which is already differentiable. However Milnor proved that S^7 has differentiable atlases which aren't equivalent via differentiable transformations (called diffeomorphisms) but which are anyway compatible with the underlying topological structure.

That means: the physics done on manifolds with an exotic structure varies from manifold to manifold. Nonequivalent exotic structures mean that the corresponding physics is different.

The big surprise came in the 1970s when out of the work of Sam Donaldson, Michael Freedman, and Clifford Taubes we learned that dimension four was very special (and dimension four is the dimension of spacetime). For, among other things:

- There are 4-dimensional manifolds which cannot be endowed with a differentiable structure.

- The 4-dimensional (hyper)plane has *uncountably many* nonequivalent differentiable structures! All of them compatible with the usual underlying topological structure.

Yes, even a simple object like a 4-plane exhibits a huge number of pathological structures!

Let there be light!

If you read Hebrew, you'll understand right away what follows. God said:

Yehi 'or. Wa yehi 'or.

If you don't, here is the translation: "let there be light. And there was light." The universe begins to exist when light is created. And light as electromagnetic waves is what we use to describe the shape of the universe.

Let's see how God's injunction is followed in the physical world.

Given an arbitrary 4-dimensional manifold M, one of the ways we have to describe it is to consider the collection of all possible submanifolds in M, that is, of all possible 3-dimensional, 2-dimensional, 1-dimensional curved spaces that fit into it. These are given by mathematical relations of a given kind, which can in turn be represented by the so-called de Rham cohomology cocycles that exist on M.[60]

These cocycles are collected into the cohomology groups of M; and they describe the shape of M through a kind of compressed code, the intersection form.

And — back to the Creator's voice: we construct the intersection form out of the so-called De Rham cohomology group $H_2(M)$, whose cocycles are electromagnetic fields — and electromagnetic fields are light! So, light is what describes the shape of the universe.

Yehi 'or.

Exotic spacetimes

Let us try to give an idea of how one constructs an exotic 4-dimensional plane, that is, an exotic R^4. First, the definition:

An exotic R^4 is homeomorphic to R^4, but not diffeomorphic to it.

(Homeomorphic: equivalent through continuous transformations; diffeomorphic: equivalent through continuous and differentiable transformations.)

In order to construct an exotic R^4, what we do is to start from a given smooth manifold and to split it in such a way that one of the pieces is the desired exotic R^4. One proceeds as follows: we start from the 4-manifold noted

$$CP^2 \# 9\overline{CP^2}. \tag{3}$$

This is the "connected sum" of the complex projective plane CP^2 and of nine copies of the same 4-manifold but with reversed orientation (that is why we placed the dash over CP^2). The corresponding intersection form is:

$$[+1] + 9[-1] \tag{4}$$

(never mind why) which precisely corresponds to the two pieces in the connected sum above.

Now we can transform that intersection form into another one (again, never mind how we've done the trick),

$$-E_8 + [-1] + [+1] \tag{5}$$

and the idea is to split the manifold coded by this intersection form into two pieces in such a way that one of the portions is an exotic R^4. Briefly, one tries to make the $[+1]$ term in the above intersection form into something spanned by a smoothly embedded sphere. But one shows that several results on intersection forms forbid it, so the sphere cannot be smooth. If we make an adequate

---

[60]Sorry for the atrocious technicalities...

restriction here, we get a smooth manifold which is topologically like R^4 but which is such that it cannot be diffeomorphic to standard R^4, as in that space the embedded sphere would have to be smooth.

This is just a very brief, sketchy description, that gives an idea of the delicate nature of the constructions required. And there is a very important property that results out of the construction above:

> Given an exotic R^4 as above, there is in that manifold a compact set which cannot be surrounded by any smooth sphere S^3.

This is very weird, as in an ordinary R^4 any compact (sort of finite) set can be surrounded by a 3-sphere which has a differentiable atlas, no matter how large is that set. One uses that idea in order to prove a mindboggling theorem by Clifford Taubes:

> R^4 has uncountably many nonequivalent differentiable structures.

The proof begins with the embedding of balls (images of the hypersphere S^3 of progressively larger radii r, r being a real number) within a family of exotic R^4s, where r codes the size of the smallest nonsmooth ball to encircle the pathological compact set. This is the easy part. The hard part is to show that all those manifolds are different (see Scorpan's book in the references for the complete argument).

What's the catch here? There is a still weirder theorem about uncountably many differentiable structures for 4-manifolds:

> Let M be a differentiable 4-manifold and let x be a point in M. Then the differentiable 4-manifold $M - \{x\}$ has uncountably many nonequivalent differentiable structures.

That is, we take a smooth manifold, cut out of it a point, and lo!, we get another manifold with uncountably many differentiable structures.

Some consequences for physics

There are several immediate consequences for physics here. A simple one: suppose that we have a system of particles that can be described by four coordinates, and suppose that the system's equations has a discontinuity at some point of coordinates (x_0, y_0, z_0, t_0). Cut out that point from the space where the system sits — but which is that space, when we have excised the singular point? Physics has always been done in local coordinates, and the global situation remains out of the description, or has to be described with the help of insufficient ad-hoc arguments.

Worse: we have mentioned forcing models in the preceding chapters. For adequate forcing constructions where we place all these objects, there will be uncountably many generic 4-exotic manifolds where to place our physics. Are they just ghostly illusions? Do exotic generic differentiable 4-manifolds have any physical meaning?

The history of mathematics and of physics shows that it is always very dangerous to eliminate solutions or alternatives just on the basis of criteria such as being sensible, natural, intuitive, and the like. So, there might be a whole new, still untouched, world behind these quite strange (at the moment) theoretical constructions.

On time

We could sum it up in an aphorism:

Time exists because the vacuum speed of light c is constant.

We'll do things backwards. First, we present our case in an admittedly technical language. Then we explain it in detail. So, don't give up, just underline the terms you don't understand and wait for the explanation.

Our slogan sounds more like a *boutade*, but it is definitely true. Consider as our basic framework that of a general relativistic spacetime. At each point on spacetime we can construct the local light cone (underline this!). Light cones exist because the local symmetry group (also this!) on spacetime is the Lorenz group, which may be seen as a reduction of the general linear group over spacetime to the Lorenz group (oops! That's a hard nut to crack!). And the Lorenz group is the linear symmetry group (without displacements) of the Maxwell–Hertz wave equations, which ask for the constant value of c.

Translation of the technical jargon. In his 1905 paper on special relativity, Einstein places the theory of electromagnetic fields above that of classical mechanics. Why is it so? He carefully anaiyzes the concept of simultaneity with the help of exchanged information about clocks in different reference systems. Information is exchanged through light ray signals in the vacuum. And electromagnetic theory provides the equation for the propagation of light rays in the vacuum; it is the wave equation.

The value of the vacuum speed of light c is determined in electromagnetic theory by the value of fundamental electromagnetic constants. So, if light is to propagate everywhere as a wave, *the speed of light in the vacuum must be a constant.*

So, if light propagates everywhere through the vacuum with constant speed c, we must accordingly change the equations of mechanics, and new phenomena (already well observed and documented) creep up, such as the Lorentz–Fitzgerald contraction and time dilation. Moreover, all physical phenomena[61] proceed with a (local) region in spacetime called the inside of the light cone.

(The light cone is the surface where we find all trajectories of light rays that emanate from a given point — called "event" — in spacetime.)

The group of transformations that preserves the form of the wave equation is the Lorentz group.[62] When we build the mathematical picture of spacetime,

---

[61] We exclude tachyons, or faster-than-light particles.

[62] Actually it's the Poincaré group, which includes translations, that we don't require here.

we get in the package a group of transformations for the objects that reside on it called the general linear group in 4 dimensions. Relativistic structures appear when we say, we don't require that big group to locally do our physics; we only need a reduction of it, the Lorentz group, at every point-event in spacetime.

This is a *broken symmetry* phenomenon, as we go down from a larger set of symmetries to a smaller one.

The interior of light cones give us the direction of time. If we piece up together all possible light cones on a spacetime we get a field that covers the whole of spacetime. It is the field of time; it may have a very crazy geometry, with swirls, bends, changes of directions. Technically, it is called a 1-foliation.

So, the fact that light propagates as a wave with constant vacuum velocity implies that the field of time exists.

If we go on with the general relativistic picture, we can go back to jargon and say that time arises when a 4-dimensional real smooth manifold is endowed with a 1-foliation, that is, with a nowhere vanishing smooth vector-field, which is the mathematical object that gives us the time arrow. We again can use a slogan:

> Spacetime is a 4-dimensional real smooth manifold with a nowhere vanishing time arrow.

And the geometry of that time arrow can be very, very complicated. Moreover the specific symmetry-breaking procedure that leads to the existence of a time arrow, that is, what mathematicians name reductions of bundle groups can have a very very complicated structure. So, we have three novelties in this picture:

- Time is described by a field. Can we say that objects become, say, temporal because they interact with the field of time?

- If time results from a broken symmetry phenomenon, then there is a situation where there is no time: it is at enormously large scales in the universe.

- What happens at the quantum level? Can we make it compatible with this large-scale picture?

So, we deal here with the question of the nature of spacetime. Gödel used the tools in Einstein's toolkit to show that time in general relativity was a much weirder animal than it seemed. We use here a few tools from Gödel's own treasure chest to try to expand even further that weirdness.

We will now restrict the scope of our enquiry. We consider here a variant of that major question. If we take that cosmic time is the time since the Big Bang, then how many models of the universe do exhibit it? Is cosmic time a typical feature of general relativistic models? Given an arbitrary model for the universe, can we prove that it has the cosmic time property?

Cosmic time? A new beast? We deal with such queries here.

Cosmic time, the Big Bang

If you are interested in physics, or in the history of the universe, then you've certainly read that according to the Big Bang picture, the universe began some 13.7 billion years ago. Now, wait a moment: if time runs at different rates in different reference systems, how can we determine with that much precision this unique age of the universe? Is there some privileged coordinate system for the universe so that we can compute over it the same age of the universe?

That works for universes whose topology of a cylinder, that is, that look like the Cartesian product $N \times R$, where R is the real line which codes the cosmic time coordinate, and N is a 3-dimensional manifold which gives the shape of the universe at each moment (in most Big Bang models, N is a 3-sphere S^3). Moreover, one has to have a *smooth, differentiable* structure that may be split into the product $N \times R$. Cylinders with that topology also have uncountably many exotic differentiable structures, but none of them has a differentiable split into a "cosmic time" coordinate R and a kind of copy of the universe at each instant, which is N. So, exotic manifolds do not allow for a Big Bang and for a cosmic time coordinate.

Now the question is: are all solutions of the Einstein gravitational equations such that they allow for a cosmic time coordinate? No. Big Bang like solutions aren't even the typical solutions among those for the Einstein equations. In fact we have:

> For a reasonable concept of "rare," the Big Bang like solutions for the Einstein gravitational equations are extremely rare.

Let's now explain the meaning of such a result. On "rare," see the next comments.

Convoluted time structures

John Archibald Wheeler suggested in the 1950s that one should take a look at the set of all possible solutions for the Einstein gravitational equations. That was done in the late 1960s and 1970s by Jerrold Marsden and his PhD student Judy Arms. Their work stems from the earlier work of René Thom (1923–2002), who invented catastrophe theory.

(A joke by Leopoldo Nachbin, the noted Brazilian expert on functional analysis: in 1981 there was a big mathematical congress in Rio, and several top-level researchers attended it; one of its hosts was Nachbin, who offered the main participants a party at his home. Nachbin was then living in a big house at Barra da Tijuca, one of Rio's upper middle-class suburbs, and his house had a beautiful garden with a swimming pool in it. Nachbin and friends were chatting by the pool when they noticed on the other side of the pool, deep in conversation, Chern, Smale and Thom. Nachbin then quipped with a malicious smile: now one of you, go to the other side of the pool, push them into it and have them drowned!

We'll finish with geometry for many decades…

Of course his companions politely refused to follow Nachbin's tongue-in-cheek advice.)

Back to general relativity. Marsden and Arms considered the space of all solutions for the Einstein equations on a given spacetime manifold.[63] The solutions with cosmic time are symmetrical solutions (in some specified sense) and Marsden and Arms showed that they organized themselves into a structure in the space of all solutions which is called a *stratification*, a kind of infinite staircase first identified by Thom in order to study these symmetrical objects.

One then sees tha the objects in the staircase-like structure are very "rare" in the space where they sit. That is to say, Big Bang-like structures are (we can say) zero probability events. That construction can be generalized and extended by elementary techniques to the space of arbitrary gauge fields — and general relativity can be seen as a gauge field — as Doria showed in 1981. (Gauge fields are the main fare in the unified theories of physics today.) The idea here is simple: the gravitational field, as any gauge field, can be represented by a "connection form," which is a matrix-like object. The different submatrices of the original matrix are related to the symmetries in the field, and they can be seen to originate a ladder-like structure that is mirrored in the space of gravitational fields.

However there is a more detailed result here, that tells us how wild is the zoo of spacetimes:

> The "typical" spacetime is exotic, without global time, and if properly axiomatized in set theory with Martin's Axiom, it is set-theoretically generic.

This results from a counting theorem: one enumerates along a line segment all possible spacetimes (there are $2^{\aleph_0}$ spacetimes). Over each one we excise a point, and get a new continuum of different inequivalent differentiable structures for the new, excised spacetime. This is enough to ensure that there are many more exotic spacetimes than the standard ones.

A brief note on "typical" and "rare." One mainly uses two different concepts in order to characterize "typical" and "rare" objects in mathematics:

- *Probability.* We endow a set with some probability measure and say that a typical object is one with 100% probability. A rare object is one with 0% probability. The interesting fact is that we may actually have a whole continuum of zero probability objects.

- *Category.* This is a topological way of describing large and small sets. A very small set is a first category set. A very large, typical, set, is a second category set.

Attention! Sometimes a second category set may have zero probability, and vice-versa. Therefore the two concepts are — weirdly? amusingly? interestingly? — inequivalent.

---

[63]Technically: the space of all Minkowskian metric tensors on some 4-manifold M.

Anything works

This happened in a GA-level talk on modern physics at one of Rio's cultural centers. The speaker briefly mentioned and tried to explain Heisenberg's uncertainty principle. A very pretty lady in the audience immediately raises her hand (she is beautiful, the speaker notices, with long smooth dark hair falling on her shoulders). May I ask a question? Yes, please do. She then says: I'm a psychoanalist. Which is the import of Heisenberg's uncertainty principle for psychoanalysis?

For a brief moment the speaker is tempted to give a standard, traditional answer: none, no relation. He hesitates, however — and gives an answer that appears at the end of this chapter.

Heisenberg's Fourth Uncertainty Relation

While Heisenberg's Uncertainty Principle for the position and momentum of a given quantum object is already very difficult to picture, there is another even stranger relation that originates in the same crucible, Heisenberg's uncertainty principle for energy and time, or Heisenberg's Fourth Relation:

$$\Delta E \, \Delta t \geq \hbar/2. \tag{6}$$

One is here showing that energy and time are irrevocably mixed. What is the meaning of that mix-up? Whence does it come from?

From classical to quantum

We'll need some mathematics here, so please be patient. The Heisenberg uncertainty principle stems from Heisenberg's commutation relation:

$$[x, p] = i\hbar, \tag{7}$$

which however mirrors the Poisson bracket relation:

$$\{x, p\} = 1 \tag{8}$$

in classical mechanics. Similarly the non-relativistic Heisenberg motion equation:

$$dA/dt = \partial A/\partial t + (-i/\hbar)[A, H], \tag{9}$$

is the counterpart of the classical motion equation

$$dA/dt = \partial A/\partial t + \{A, H\}. \tag{10}$$

Here H is the so-called Hamiltonian function. If it doesn't explicitly depend on time t, the Hamiltonian is the system's total energy. Since classical mechanics, the Hamiltonian kind of codes the way the system correctly evolves along time.

It is a kind of program for time evolution. So, time & energy are inextricably mixed up in physics, both classical and quantum.

Moreover, if you have some training in general relativity, you look at both the equations above and notice that they are parallel transport equations, where H stands in the place of the gravitational affine connection. Actually we can formulate a theory of such motion equations as what is technically called a theory of a partial connection on a manifold.

Heisenberg's fourth relation — the other three tell about the uncertainty between position x and momentum p — suggests a rather radical theory of the birth of the universe: it is a quantum fluctuation of the vacuum. For one such fluctuation, with a very small value of ΔE, the corresponding time duration of the fluctuation Δt might be extremely large.

Quite beautiful: we all would be nothing but a random oscilation of the quantum vacuum. No Big Bang, no big splash, just an oscillation of the vacuum. As in *Macbeth*'s lines:

> ...*a walking shadow, a poor player*
> *who struts and frets his hour upon the stage*
> *and then is heard no more*...

Wiener integrals, Feynman integrals and the Multiverse

In Chapter 2, we considered sums over all programs like the summation that is used to define the halting probability Ω. Now we'd like to go from sums over all paths (Wiener) to sums over all histories (Feynman) to sums over all geometries (Wheeler) to sums over all universes (Tegmark)!

The Wiener measure is a probability on the space of all continuous functions and is basically just brownian motion. Here is a beautiful application of Wiener measure and the corresponding Wiener integral: We will use the Wiener integral to solve an important partial differential equation. Given boundary conditions, we get the solution to Laplace's equation (value at point x = average of value on infinitesimal sphere centered on x) by taking the average value at the point on the boundary that you eventually hit by taking a random walk starting at the point x. For more on this, see Mark Kac, *Probability and Related Topics in Physical Sciences*.

Now let's turn to the Feynman path integral = path democracy = sum over all histories = sum over all trajectories. There is a

$$\exp\left[\frac{i}{\hbar} \int \mathcal{L}\, dt\right] \tag{11}$$

weight on each path. Here $\mathcal{L}$ = the Lagrangian and is the difference between the kinetic and the potential energy at each point on the path. The actual classical path minimizes the integral of the Lagrangian $\mathcal{L}$ over the entire path; this is the famous *principle of least action*.[64] In quantum mechanics you do not have a single trajectory; many paths may contribute to the final answer, to what you get by performing a measurement on the system.

Feynman's basic idea is that a physical system simultaneously performs a time evolution over all possible histories, which interfere constructively and destructively according to the quantum amplitude (complex-valued probability) $e^{iS/\hbar}$ associated with each path, which depends on the action $S = \int \mathcal{L}\, dt$. In the classical world, all but one of these paths cancel out due to quantum interference; but in the quantum world this fails to occur.

This also explains the basic idea of *quantum computation*, which is to exploit the *quantum parallelism* revealed by the Feynman path integral. A quantum computer will simultaneously perform a great many, ideally exponentially many, computations in parallel.[65]

For more on the Feynman integral and the principle of least action, see Feynman, *QED: The Strange Theory of Light and Matter*, *The Feynman Lectures on Physics*, and Feynman and Hibbs, *Quantum Mechanics and Path Integrals*.

The Wiener integral is mathematically rigorous. Physicists love the Feynman path integral formulation of quantum field theory and Feynman integrals are tremendously useful, but it has not yet been possible to formulate Feynman integrals in a mathematically rigorous fashion. Now let's look at some stuff that is even more weird.

Inspired by the Feynman path integral formulation of quantum mechanics, the huge book Wheeler et al., *Gravitation*, has a chapter on sums over all space-time geometries and on quantum fluctuations in gravitational fields. By analogy with classical electrodynamics, Wheeler et al. propose calling this new field *geometrodynamics*. Unfortunately, although the physical intuition is compelling, the field of quantum gravity remains largely *terra incognita*, in spite of decades of work on quantum gravity since *Gravitation* was published.

Now let's go from geometrodynamics, to Susskind's cosmic landscape, to the multiverse celebrated by Deutsch, Tegmark, Vilenkin and others. First see Leonard Susskind, *The Cosmic Landscape*, for the "landscape" of all possible string theories, a relatively down-to-earth kind of multiverse.[66] And for much more extreme multiverses, see David Deutsch, *The Fabric of Reality*, and the Max Tegmark article in the big John Barrow et al., John Wheeler festschrift published by Cambridge University Press.

---

[64]The principle of least action is often attributed to Maupertuis but is in fact actually due to Leibniz.

[65]Nobody knows yet if such computers will actually work, but there are already useful technologies emerging from the attempt to build quantum computers. As Raymond Laflamme, the head of the Waterloo, Canada, Institute for Quantum Computing explains, this is a win-win situation. For if quantum computers are so sensitive to perturbations and noise that they decohere and fail to work, then instead of getting working quantum computers what you get is extremely sensitive new measurement technology. Either way you win.

[66]String theory is an attempt to construct a theory of quantum gravity by replacing point particles by strings in order to avoid all the divergencies, all the calculations that give infinite answers. Basically there is a minimum distance scale given by the width of a string.

According to Tegmark and others, the laws of this particular universe are of no fundamental interest; they are just our postal address in the multiverse, in the space of all possible universes! And who cares about a postal address?! What is fundamental is the multiverse, the ensemble of all possible universes. That is a conceptual structure worthy of study, not the particular laws of this particular, uninteresting universe!

Here is another way that Tegmark explains this: The infinite set $\{1, 2, 3, \ldots\}$ of all positive integers is very simple to calculate, but some positive integers, e.g., 9353942376348762376384767, are very complicated, they require arbitrarily large programs. Similarly, according to Tegmark the space, the ensemble, of *all possible* physical universes is simpler than having to specify *individual* universes.

When considering the multiverse of all possible universes, we encounter this fundamental

Measure problem: *What weight should we put on each possible universe?*[67]

If following Tegmark the multiverse consists of all possible mathematical structures, then the measure problem is an extremely difficult problem, like attempting to rigorously define Borel's know-it-all oracle number. But if each universe = software, then there is no problem, none at all! Following Chapter 2, the weight of the universe defined by the program p is just $2^{-|p|}$, 1 over 2 raised to the size in bits of p. This is what we get by replacing God the mathematician with God the computer programmer: an easy solution to the problem of what measure, what *a priori* probability, to associate with a given universe p.

In spite of these suggestions to give up on our own universe and instead consider the multiverse of all possible universes, Stephen Wolfram is actually conducting a systematic computer search for the laws of this universe, along the lines described in Chapter 9 of *A New Kind of Science*.

Envoi: On Eternity and Beyond

Now a vignette from a series of lectures given by Valentine Bargmann, Einstein's coworker, in the early 1980s. Bargmann lectured on general relativity at Rochester in January 1980. Bargmann was housed in a corner office in the 4th floor of the math sciences building at the University of Rochester's wooded campus, and one of the students who attended the lectures office was the next door office. That student was then a junior postdoc with interests in the geometry of gauge fields, and his PhD thesis dealt with the Bargmann–Wigner equations (in a way that Bargmann detested, by the way), but nevertheless the boy would nearly corner the old man for a chat, to which Bargmann would always graciously oblige.

Bargmann told the student about several interesting developments from the heroic times of the birth and early growth of general relativity and quantum

---

[67]We thank Alexander Vilenkin, author of *Many Worlds in One*, for telling us about the measure problem.

mechanics. For example, he mentioned the need to look into several loose ends, like Felix Ehrenhafts' subelectrons, and his uncanny magnetic monopoles. Ehrenhaft was a great experimentalist, Bargmann would stress. Go to the library, look at the issues of *The Physical Review* in the forties (1940s). The Ehrenhaft papers only appear at the end of the issues, as unrefereed letters, as no one would accept a paper on the subjects he exposed, but they would anyway allow them to be published as letters.

Once the boy asked Bargmann about the metaphysics that were said to be espoused by John Archibald Wheeler and his group. Bargmann was cryptic: it looks like a kind of religion, he said.

Loose ends, forays into untrodden woods — like the title of another of Heidegger's books, *Holzwege*, forays into the woods. Unexplored inroads?

Carl Gustav Jung (1875–1961) was Freud's favorite disciple until they parted ways — bitterly — in the 1910s. Jung's ideas were, and still are, believed to belong to the lunatic fringe, or nearly so, even if his contributions to psychiatry were considerable. In fact, we may even say that Jung was a kind of Nikola Tesla in psychiatry and psychoanalysis. Jung claimed that at its very bottom our personal unconscious minds merge into a vast Unconscious which is shared by everybody and everything in the universe. That concept, which was modelled after Jung's personal understanding of the Eastern mystical lore, implies that in that Unconscious time doesn't exist.

Sheer wild speculation? Well, Jung was close to two great 20th century physicists, Wolfgang Pauli (of the Pauli Exclusion Principle, which explains the solidity of matter, among other things, 1900–1958) and Pauli's student Markus Fierz. Jung and Pauli even wrote a book together, *Synchronicity*, on Jung's mysterious acausal principle that underpins everything according to the Junguian viewpoint. (Pauli's contribution to that book is, by the way, quite well-behaved.) We get a flavor of these Jungian ideas in some ideas that are said to have been privately espoused by Wheeler's disciples in the 1960s.

This is what the lecturer we've mentioned before (page 125) answered the beautiful psychoanalyst. He carefully stressed that these were highly tentative, speculative ideas. (We actually don't know if the lecturer was only trying to impress the lady, of course, but someone noticed that the lady had a mischievous smile on her lips.) And he concluded:

This is the stuff that dreams are made of.

Powerful dreams they are.

References

[1] — *Minds and Machines* **13** (2003).

[2] V. I. Arnold et al., *Mathematics: Frontiers and Perspectives*, AMS/IMU (2000).

[3] A. I. Arruda et al., *Mathematical Logic in Latin America*, North-Holland (1980).

[4] T. Asselmeyer-Maluga and C. H. Brans, *Exotic Smoothness and Physics*, World Scientific (2007).

[5] V. Bargmann, "Lectures on general relativity," xeroxed lecture notes, Univ. of Rochester, NY (1980).

[6] V. Bargmann, memories of Gödel and Einstein, personal communication to F. A. Doria, Rochester, NY (1980).

[7] J. Barrow, *Impossibility: The Limits of Science and the Science of Limits*, Oxford (1998).

[8] J. Barrow et al., *Science and Ultimate Reality*, Cambridge University Press (2004).

[9] T. Baker, J. Gill and R. Solovay, "Relativizations of the $P =?NP$ question," *SIAM J. Comp.* **4**, 431–442 (1975).

[10] L. Beklemishev, "Provability and reflection," Lecture Notes for ESSLLI'97 (1997).

[11] S. Ben-David and S. Halevi, "On the independence of $P\ vs.\ NP$, Technical Report # 699, Technion (1991).

[12] P. Billingsley, *Ergodic Theory and Information*, Wiley (1965).

[13] L. Blum, F. Cucker, M. Shub and S. Smale, *Complexity and Real Computation*, Springer (1998).

[14] É. Borel, *Leçons sur la théorie des fonctions*, Gauthier-Villars (1950), Gabay (2003), pp. 275.

[15] É. Borel, *Les Nombres inaccessibles*, Gauthier-Villars (1952), p. 21.

[16] É. Borel, *Space and Time*, Dover (1960), pp. 212–214.

[17] S. Bringsjord and M. Zensen, *Superminds*, Springer (2005).

[18] A. Burks, *Essays on Cellular Automata*, University of Illinois Press (1970).

[19] C. Calude, *Randomness and Complexity, from Leibniz to Chaitin*, World Scientific (2007).

[20] W. A. Carnielli and F. A. Doria, "Is computer science logic dependent?" in C. Dégremont et al., *Dialogues, Logics and other Strange Things: Essays in Honor of Shahid Rahman*, College Publications (2008).

[21] P. Cassou-Noguès, *Les Démons de Gödel*, Seuil (2007).

[22] G. J. Chaitin, "On the length of programs for computing finite binary sequences," *Journal of the ACM*, **13**, 547–569 (1966).

[23] G. J. Chaitin, "Information-theoretic limitations of formal systems," *Journal of the ACM*, **21**, 403–424 (1974).

[24] G. J. Chaitin, *Algorithmic Information Theory*, Cambridge University Press (1987).

[25] G. J. Chaitin, *Information Randomness and Incompleteness*, World Scientific (1987).

[26] G. J. Chaitin, "Computing the Busy Beaver function," in T. M. Cover and B. Gopinath, *Problems in Communication and Computation*, Springer (1987).

[27] G. J. Chaitin, *The Limits of Mathematics*, Springer (1998).

[28] G. J. Chaitin, *Exploring Randomness*, Springer (2001).

[29] G. J. Chaitin, *Meta Math!*, Pantheon (2005).

[30] G. J. Chaitin, *Mathematics, Complexity and Philosophy*, Midas, in press (2010).

[31] E. C. Cherry, *On Human Communication*, MIT Press (1966).

[32] A. Church, "An unsolvable problem of elementary number theory," *Amer. J. Math* **58**, 345–363 (1936).

[33] E. Codd, *Cellular Automata*, Academic Press (1968).

[34] P. J. Cohen, *Set Theory and the Continuum Hypothesis*, Addison-Wesley (1966).

[35] B. J. Coleman, "The Church–Turing Thesis," *The Stanford Encyclopedia of Philosophy*, http://plato.stanford.edu/entries/church-turing (2002).

[36] B. J. Coleman and R. Sylvan, "Beyond the universal Turing machine," *Australasian J. Philosophy* **77**, 46–66 (1999).

[37] N. C. A. da Costa, "Sur un système inconsistant de théorie des ensembles," *C. R. Acad. Sci. de Paris* **258**, 3144–3147 (1964).

[38] N. C. A. da Costa, "Sur une hiérarchie de systèmes formels," *C. R. Acad. Sci. de Paris* **259**, 2943–2945 (1964).

[39] N. C. A. da Costa, "On the theory of inconsistent formal systems," *Notre Dame J. of Formal Logic* **11**, 497–510 (1974).

[40] N. C. A. da Costa, *Logiques Classiques et Non-Classiques*, Masson (1997).

[41] N. C. A. da Costa and O. Bueno, "On paraconsistency: towards a tentative interpretation," *Theoria* **16**, 119–145 (2001).

[42] N. C. A. da Costa and O. Bueno, "Quasi-truth, paraconsistency, and the foundations of science," *Synthèse* **154**, 283–289 (2007).

[43] N. C. A. da Costa and R. Chuaqui, "On Suppes' set-theoretical predicates," *Erkenntnis* **29**, 95 (1988).

[44] N. C. A. da Costa and F. A. Doria, "Undecidability and incompletenetess in classical mechanics," *Int. J. Theor. Phys.* **30**, 1041 (1991).

[45] N. C. A. da Costa and F. A. Doria, "Classical physics and Penrose's Thesis," *Found. Phys. Letters* **4**, 363 (1991).

[46] N. C. A. da Costa and F. A. Doria, "Gödel incompletenetess in analysis, with an application to the forecasting problem in the social sciences," *Philosophia Naturalis* **31**, 1 (1994).

[47] N. C. A. da Costa and F. A. Doria, "Suppes predicates and the construction of unsolvable problems in the axiomatized sciences," P. Humphreys, ed., *Patrick Suppes, Scientific Philosopher*, II, Kluwer (1994).

[48] N. C. A. da Costa and F. A. Doria, "Variations on an Original Theme," in J. L. Casti and A. Karlqvist, *Boundaries and Barriers*, Addison-Wesley (1996).

[49] N. C. A. da Costa and F. A. Doria, "Structures, Suppes predicates, and Boolean-valued models in physics," in P. Bystrov and V. Sadovsky, eds., *Philosophical Logic and Logical Philosophy — Essays in Honor of Vladimir A. Smirnov*, Synthèse Library, Kluwer (1996).

[50] N. C. A. da Costa and F. A. Doria, "Consequences of an exotic formulation for $P = NP$," *Applied Mathematics and Computation* **145**, 655–665 (2003); also "Addendum," *Applied Mathematics and Computation* **172**, 1364–1367 (2006).

[51] N. C. A. da Costa and F. A. Doria, "Computing the future," in K. V. Velupillai, ed., *Computability, Complexity and Constructivity in Economic Analysis*, 15–50, Blackwell (2005).

[52] N. C. A. da Costa and F. A. Doria, *On the Foundations of Science, I,* e-papers/GAE–PEP–COPPE (2008).

[53] N. C. A. da Costa and F. A. Doria, "Hypotheses that imply the independence of $P = NP$ from strong axiomatic systems," in S. Zambelli, ed., *Computable, Constructive and Behavioural Economic Dynamics,* 79–102, Routledge (2010).

[54] N. C. A. da Costa, F. A. Doria and J. A. de Barros, "A Suppes predicate for general relativity and set-theoretically generic spacetimes," *International Journal of Theoretical Physics,* **29**, 935–961 (1990).

[55] N. C. A. da Costa, F. A. Doria, E. Bir, "On the metamathematics of the P vs. NP question," *Applied Math. and Computation* **189**, 1223–1240 (2007).

[56] N. C. A. da Costa and S. French, *Science and Partial Truth,* Oxford (2003).

[57] N. C. A. da Costa and N. Grana, *Il Ricupero dell'Inconsistenza,* L'Orientale Editrice (2009).

[58] N. C. A. da Costa, D. Krause and O. Bueno, "Paraconsistent logics and paraconsistency," in D. Jacquette, ed., *Handbook of the Philosophy of Science: Philosophy of Logic,* 791–911, Elsevier (2007).

[59] A. Coudert, *Leibniz and the Kabbalah,* Kluwer (1995).

[60] D. Deutsch, *The Fabric of Reality,* Basic Books (1997).

[61] G. Debreu, *Mathematical Economics: Twenty Papers of Gerard Debreu,* Cambridge (1983).

[62] F. A. Doria, "Informal vs. formal mathematics," *Synthèse* **154**, 401–415 (2007).

[63] F. A. Doria, *Chaos, Computers, Games and Time,* e-papers/COPPE-PEP (2011).

[64] F. A. Doria and J. F. Costa, editors, "Hypercomputation," special issue of *Applied Mathematics and Computation, Applied Mathematics and Computation* **178** (2006).

[65] F. A. Doria and M. Doria, "On formal structures in general relativity," *Philosophia Naturalis* **46**, 115–132 (2009).

[66] F. A. Doria and M. Doria, "Einstein and Gödel: on time structures in general relativity," in D. Krause and A. A. Leite Videira, *Festschrift in Honor of Newton da Costa,* Boston Studies in the Philosophy of Science, Springer (2011).

[67] F. Dyson, *The Sun, the Genome and the Internet,* Oxford University Press (1999).

[68] F. Dyson, *A Many-Colored Glass*, University of Virginia Press (2007).

[69] U. Eco, *The Search for the Perfect Language*, Blackwell (1995).

[70] R. L. Epstein, W. A. Carnielli, *Computability: Computable Functions, Logic, and the Foundations of Mathematics*, Wadsworth Publishing, 2nd ed. (1999).

[71] S. Feferman, "Arithmetizations of metamathematics in a general setting," *Fund. Mathematica* **49**, 35–92 (1960).

[72] S. Feferman, "Transfinite recursive progressions of axiomatic theories," *J. Symbolic Logic* **27**, 259–316 (1962).

[73] S. Feferman et al., eds., *Kurt Gödel, Collected Works*, I, Oxford (1986).

[74] R. Feynman, *QED: The Strange Theory of Light and Matter*, Princeton University Press (1985).

[75] R. Feynman et al., *The Feynman Lectures on Physics*, Addison-Wesley (1964).

[76] R. P. Feynman and A. R. Hibbs, *Quantum Mechanics and Path Integrals*, McGraw-Hill (1965).

[77] T. Franzen, "Transfinite progressions: a second look at completeness," *Bull. Symbolic Logic* **10**, 367-389 (2004).

[78] E. Fredkin, *http://www.digitalphilosophy.org*.

[79] J. Fresán, "Review of *Pythagoras' Revenge*," *AMS Notices* **57**, 637–638 (2010).

[80] D. Gabbay, P. Thaggard and J. Woods, eds., *Handbook of the Philosophy of Science*, **5**: *Philosophy of Logic*, Elsevier (2006).

[81] G. Gentzen, "Die Widerspruchsfreiheit der reinen Zahlentheorie," *Math. Ann.* **112**, 493–565.

[82] K. Gödel, "Über formal unentscheidbare Sätze der Principia Mathematica und verwandter Systeme I," *Monatsh. Math. Phys* **38** 173–198 (1931).

[83] K. Gödel, "An example of a new type of cosmological solution of Einstein's field equations of gravitation," *Rev. Mod. Phys.*, **21**, 447–450 (1949).

[84] R. Goldstein, *Incompleteness: The Proof and Paradox of Kurt Gödel*, Norton (2006).

[85] R. Gompf, A. I. Stipsicz, *4-Manifolds and Kirby Calculus*, AMS (1999).

[86] C. Goodman-Strauss, "Can't decide? Undecide!," *AMS Notices* **57**, 343–356 (2010).

[87] S. Gould, *Wonderful Life*, Norton (1989).

[88] N. Grana, *Epistemologia della Matematica*, L'Orientale Editrice (2001).

[89] J. Hartmanis and J. Hopcroft, "Independence results in computer science," SIGACT News **13** (1976).

[90] M. Heidegger, *Einführung in die Metaphysik*, Max Niemeyer (1953).

[91] M. Heidegger, *Wegmarken*, Vittorio Klostermann (1967).

[92] F. Hoyle, *Ossian's Ride*, Harper (1959).

[93] J. Horgan, *The End of Science*, Addison-Wesley (1996).

[94] M. Kac, *Probability and Related Topics in Physical Sciences*, Interscience (1959).

[95] S. C. Kleene, "General recursive functions of natural numbers," *Math. Ann.* **112**, 727–742 (1936).

[96] S. C. Kleene, "General recursive functions of natural numbers," *Math. Annalen* **112**, 727–742 (1935/6).

[97] S. C. Kleene, "λ-definability and recursiveness," *Duke Math. J.* **2**, 340–353 (1936).

[98] S. C. Kleene, *Introduction to Metamathematics*, Van Nostrand (1952).

[99] S. C. Kleene, *Mathematical Logic*, John Wiley (1967).

[100] G. T. Kneebone, *Mathematical Logic and the Foundations of Mathematics*, Van Nostrand (1963).

[101] N. Koblitz and A. Menezes, "The brave new world of bodacious assumptions in cryptography," *AMS Notices* **57**, 357–365 (2010).

[102] R. Koppl, "Thinking impossible things: A Review Essay," *Journal of Economic Behavior and Organization* **66**, 837–847 (2008).

[103] R. Koppl, "Computable Entrepreneurship," *Entrepreneurship Theory and Practice* **32**, 919–926 (2008).

[104] R. Koppl, "Complexity and Austrian Economics," in J. Barkley Rosser, Jr, ed., *Handbook on Complexity Research*, Edward Elgar (2009).

[105] R. Koppl, "Some epistemological implications of economic complexity," *Journal of Economic Behavior and Organization* **76**, 859–872 (2010).

[106] R. Koppl and J. Barkley Rosser, Jr., "All that I have to say has already crossed your mind," *Metroeconomica* **53**, 339–360 (2002).

[107] G. Kreisel, "A notion of mechanistic theory," in P. Suppes, ed., *Logic and Probability in Quantum Mechanics*, D. Reidel (1976).

[108] H. W. Kuhn and S. Nasar, *The Essential John Nash*, Princeton (2002).

[109] K. Kunen, "A Ramsey theorem in Boyer–Moore logic," *J. Automated Reasoning* **15** 217–224 (1995).

[110] C. Lanczos, "Über eine stationäre Kosmologie im Sinne der Einsteinschen Gravitationstheorie," *Zeitschrift für Physik*, 21, 73–110 (1924).

[111] T. Leiber, *Kosmos Kausalität und Chaos*, Ergon Verlag (1996).

[112] G. W. Leibniz, *Discours de métaphysique, suivi de Monadologie*, Gallimard (1995).

[113] M. Machtey and P. Young, *An Introduction to the General Theory of Algorithms*, North-Holland (1979).

[114] K. Menger, "Das Botenproblem," in K. Menger, ed., *Ergebnisse eines Mathematischen Kolloquiums* 2, 11–12 (1932).

[115] I. Mikenberg, N. C. A. da Costa, R. Chuaqui, "Pragmatic truth and approximation to truth," *British J. for the Philosophy of Science*, **40**, 333–358 (1989).

[116] D. Miller, *Critical Rationalism*, Open Court (1994).

[117] E. Nagel and J. Newman, *Gödel's Proof*, Routledge & Kegan Paul (1964).

[118] J. von Neumann, *Theory of Self-Reproducing Automata*, University of Illinois Press (1966).

[119] M. O'Donnell, "A programming language theorem which is independent of Peano Arithmetic," *Proc. 11th Ann. ACM Symp. on the Theory of Computation*, 176–188 (1979).

[120] E. Post, "Recursively enumerable sets of positive integers and their decision problems," *Bull. Amer. Math. Society*, **50**, 284–316 (1944).

[121] T. Radò, "On non-computable functions, *Bell System Technical J.*, **41**,877884 (1962).

[122] E. Regis, *Who got Einstein's Office?*, Addison–Wesley (1987).

[123] P. Ribenboim, *The Book of Prime Number Records*, Springer (1989).

[124] H. G. Rice, "Classes of recursively enumerable sets and their decision problems," *Transactions of the AMS*, **74**, 358–366 (1953).

[125] K. Riezler, "Das homerische Gleichnis und der Anfang der Philosophie," *Die. Antike* **12**, 253–271 (1936).

[126] H. Rogers Jr., *Theory of Recursive Functions and Effective Computability*, MIT Press (1967).

[127] A. Sangalli, *Pythagoras' Revenge*, Princeton University Press (2009).

[128] B. Scarpellini, "Two undecidable problems of analysis," translation, *Minds and Machines* **13**, 49–77 (2003).

[129] B. Scarpellini, "Comments to 'Two undecidable problems of analysis,'" *Minds and Machines* **13**, 79–85 (2003).

[130] A. Scorpan, *The Wild World of 4-Manifolds*, AMS (2005).

[131] D. S. Scott, "Lectures on a mathematical theory of computation," in M. Broy and G. Schmidt, eds., *Theoretical Foundations of Programming Methodology*, D. Reidel (1982).

[132] C. E. Shannon, "A mathematical theory of communication," *Bell Systems Tech. J.* **27**, 379–423 and 623–656 (1948).

[133] N. Shubin, *Your Inner Fish*, Pantheon (2008).

[134] H. Siegelmann, *Neural Networks and Analog Computation: Beyond the Turing Limit*, Birkhäuser (1999).

[135] L. Smolin, *Three Roads to Quantum Gravity*, Basic Books (2001).

[136] , I. Stewart, "Deciding the undecidable," *Nature* **352**, 664–665 (1991).

[137] I. Stewart, *From Here to Infinity*, Oxford (1996).

[138] L. Susskind, *The Cosmic Landscape*, Little, Brown (2006).

[139] A. Syropoulos, *Hypercomputation*, Springer (2008).

[140] M. Tsuji, N. C. A. da Costa and F. A. Doria, "The incompleteness of theories of games," *J. Phil. Logic* **27** 553–563 (1998).

[141] A. Turing, "On computable numbers, with an application to the Entscheidungsproblem," *Proc. London Math. Society* **42**, 230–265 (1937).

[142] A. Turing, "System of logic based on ordinals," *Proc. London Math. Society* series 2 **45**, 161–228 (1939).

[143] T. Tymoczko, *New Directions in the Philosophy of Mathematics*, Princeton University Press (1998).

[144] K. V. Velupillai, "Richard Goodwin, 1913–1996," *The Economic Journal* **108**, 1436–1449 (1998).

[145] K. V. Velupillai, ed., *Computability, Complexity and Constructivity in Economic Analysis*, Blackwell (2005).

[146] K. V. Velupillai, "A computable economist's perspective on computational complexity," in J. Barkley Rosser Jr., ed., *The Handbook of Complexity Research*, Ch. 4, 36–83, Edward Elgar Publishing Ltd. (2009).

[147] K. V. Velupillai, *Computable Foundations for Economics*, Routledge (2010).

[148] K. V. Velupillai, "Towards an algorithmic revolution in economic theory," *J. Economic Surveys* **25**, 401–430 (2011).

[149] K. V. Velupillai, e-mail messages do F. A. Doria (2003 onwards).

[150] K. V. Velupillai, S. Zambelli and S. Kinsella, *The Elgar Companion to Computable Economics*, Edward Elgar, Cheltenham (2011).

[151] A. Vilenkin, *Many Worlds in One*, Hill and Wang (2006).

[152] H. Weyl, *The Open World*, Yale University Press (1932).

[153] J. Wheeler et al., *Gravitation*, Freeman (1973).

[154] S. Wolfram, *A New Kind of Science*, Wolfram Media (2002).

[155] W. H. Woodin, "The continuum hypothesis, Part I," *AMS Notices* **48**, 567–576 (2001).

[156] P. Yourgrau, *A World without Time*, Basic Books (2005).

[157] S. Zambelli, "Computable and constructive dynamics ...," in S. Zambelli, ed., *Computable, Constructive and Behavioural Economic Dynamics*, 1–12, Routledge (2010).

[158] K. Zuse, *Rechnender Raum* (Calculating Space), Vieweg (1969).

Milton Keynes UK
Ingram Content Group UK Ltd.
UKHW040052071024
449327UK00019B/498